Photovoltaic Power for Europe

Solar Energy R&D in the European Community

Series C,

Photovoltaic Power Generation

Volume 2

Publication arrangements: D. NICOLAY

Solar Energy R&D
in the European Community

Series C   Volume 2
Photovoltaic Power Generation

Photovoltaic Power for Europe

An Assessment Study

by

MICHAEL R. STARR
Sir William Halcrow and Partners, Swindon, England

and

W. PALZ
Commission of the European Communities

EUR 8366

D. REIDEL PUBLISHING COMPANY

Dordrecht, Holland / Boston, U.S.A. / London, England

for the Commission of the European Communities

Library of Congress Cataloging in Publication Data

Starr, Michael, 1941–
    Photovoltaic power for Europe.
    (Solar energy R&D in the European community. Series C. ; v. 2)
    Bibliography: p.
    1. Photovoltaic power generation–Europe. 2. Solar energy
industries–Europe. I. Palz, Wolfgang. II. Commission of the
European Communities. III. Title. IV. Series.
TK2960.S73   1983   333.79'23   82-25008
ISBN 90-277-1556-4

This assessment study was prepared at the request of the
Commission of the European Communities,
Directorate-General Science, Research and Development, Brussels

Publication arrangements by
Commission of the European Communities
Directorate-General Information Market and Innovation, Luxembourg

EUR 8366
Copyright © 1983, ECSC, EEC, EAEC, Brussels and Luxembourg

Published by D. Reidel Publishing Company
P.O. Box 17, 3300 AA Dordrecht, Holland

Sold and distributed in the U.S.A. and Canada
by Kluwer Boston Inc.,
190 Old Derby Street, Hingham, MA 02043, U.S.A.

In all other countries, sold and distributed
by Kluwer Academic Publishers Group,
P.O. Box 322, 3300 AH Dordrecht, Holland

D. Reidel Publishing Company is a member of the Kluwer Group

Solar energy is seen by many as an important renewable energy resource which will be increasingly used in the future to help meet the world's growing energy requirements. Photovoltaic devices, which convert light directly into electricity, provide a particularly attractive and promising method of solar energy utilization. Photovoltaic systems received some attention in the 1950's for powering small specialist equipment in remote areas, but the main thrust of development came in the 1960's when photovoltaic generators first came into regular use for powering space satellites. Since the oil crisis in the early 1970's, efforts have been directed worldwide to develop and apply this technology for terrestrial applications.

Until now, total investment into photovoltaics since the mid-1970's has exceeded $1 billion; most spending has been in the United States. Besides tremendous investments of the industries themselves, in particular the oil companies, the US Government from 1974 to the late 1970's has increased its R&D budget from $1 million to more than $100 million.

From 1972 onwards many appraisals of photovoltaics for terrestrial applications have been performed in the United States to evaluate very precisely possible cost reductions and establish price goals for photovoltaics as a large-scale power source. Generally speaking, these assessment studies concluded that photovoltaics would eventually become competitive with other conventional types of power plants, probably in the 1990's rather than in the late 1980's. Outside the photovoltaic community, these projections have been viewed with scepticism but looking back today, one has to admit that most intermediate goals of the US photovoltaic programme have been well achieved.

For Europe, to our knowledge, no comprehensive assessment study for photovoltaics has been published until now, leaving aside certain attempts made at a very early stage (eg in France by Centre National d'Etudes Spatiales who prepared a photovoltaic asessment for the French Ministry Of Industry in 1974).

18 months ago the Commission felt that the time was ripe to start a comprehensive assessment study for the European Community. Not only had the structure and commitment of European industry towards photovoltaics developed tremendously in the last few years and major new investments been decided, but also the production and market conditions in Europe are so different from those in the United States that a specific approach to the problem from a European point of view was overdue.

Whilst a West German manufacturer is still the world's leading supplier of silicon material, a key component of photovoltaics, the share of the world's total market for photovoltaics taken by European suppliers is still small but steadily increasing. The bulk of the investments for improving cell technology and establishing large automated production lines in Europe have still to be made. Even when admitting that photovoltaics may supply one day a significant share of Europe's electricity needs, there are many doubts about the prospects for intermediate markets.

In this book an attempt has been made to respond to the many open questions and uncertainties associated with photovoltaics today. As far as this is possible, the approach taken was objective and realistic. Experts from almost all member countries of the European Community have been involved in the attempt to gain a balanced view. Industry's views have been taken as a prime input to the study.

So far, much of the photovoltaic research and development work undertaken in Europe has been co-ordinated and supported by the Commission of the European Communities. This report has been prepared for the Commission by Sir William Halcrow and Partners, Consulting Engineers, under Contract No. ESC-P-049-81-UK (H), in order to provide an independent assessment of the potential of photovoltaics over next twenty years, as an energy resource for Europe, as an export for European industry and as part of European aid to developing countries. The report includes a review of the future trends in prices and system sales. The social, legal and environmental implications of photovoltaic power, if implemented on a large scale, are discussed and the present position and prospects of the emerging European photovoltaic industry are reviewed. Suggestions for future research and development priorities are presented, together with proposals for a co-ordinated plan for implementing photovoltaics as a new energy resource of major significance in Europe.

The report was prepared by M R Starr of Sir William Halcrow and Partners under the direction of W Palz, Head of the Commission's R&D Programme for Solar Energy.

Acknowledgements

Special thanks are due to the following for their helpful contributions:

Y Chevalier	COMES, France
K Krebs	Joint Research Centre, Ispra, Italy
H Macomber	Monegon Limited, USA
R Van Overstraeten	KU Leuven, Belgium
S Pizzini	Heliosil, Italy
A Taschini	ENEL, Italy
F C Treble	Consultant, UK
T Vyverman	National Programme for Energy Research and Development, Belgium

The Commission's photovoltaic advisory panel known as Expert Group C provided much useful information and advice, in particular by reviewing the report whilst in draft form. In addition to the individuals mentioned above, valuable comments were received from W H Bloss (FR Germany), A Dollery (UK), Ph Dumont (Belgium), E Fabre (France), A Higgins (Ireland), A H M Kipperman (The Netherlands), C N Kourogenis (Greece), C Malaguti (Italy), R P M Sonneville (The Netherlands), P Wolfe (UK), G T Wrixon (Ireland) and K Zollenkopf (FR Germany). Thanks are also due to the staff of Intermediate Technology Power Limited (UK) for their comments and advice.

Dr Starr had discussions with all the principal manufacturers of photovoltaic materials and components in Europe in the course of gathering material for the report, including the following whose help and hospitality are gratefully acknowledged:

AEG–Telefunken   (FR Germany)
Fabricable/IDE   (Belgium)
France Photon   (France)
Holec   (The Netherlands)
Lucas BP Solar Systems   (UK)
Photowatt   (France)
Pragma   (Italy)
Siemens   (FR Germany)
Wacker–Chemitronic   (FR Germany)

The information and comments received from many other persons and organizations, too numerous to mention individually, are gratefully acknowledged.

NOTE ON CURRENCY UNITS AND PRICES

In this report, data with monetary values are generally expressed in European Currency Units, ECU. The ECU is a 'basket' unit, based on a certain quantity of each Community member currency, weighted on the basis of average GNP, the intra-Community trade and other factors. The monetary parities between the ECU and individual currencies vary continuously and for the purposes of this report the following exchange rates have been used:

```
ECU 1.00 =    44.97   Belgium Franc
               8.23   Danish Krone
               2.33   German D-mark
               6.61   French Franc
               2.58   Dutch Guilder
              66.75   Greek Drachma
               0.69   Irish Punt
            1350.27   Italian Lira
               0.55   UK Pound
               0.94   US Dollar
             247.75   Japanese Yen
```

(Based on September 1982 exchange rates)

All prices given for photovoltaic modules and systems, including forecasts, are given in 1980 currency values unless otherwise stated, since 1980 is the base year used by many authorities for comparing photovoltaic prices.

C O N T E N T S

CHAPTER SEVEN - CONCLUSIONS AND RECOMMENDATIONS

CHAPTER ONE - INTRODUCTION

1.1 Solar Energy

One of the greatest challenges facing the world today is the development of new and renewable sources of energy that can supplement and, where appropriate, replace the diminishing resources of fossil fuels. Solar energy, used in one form or another, is clearly one of the most promising prospects since every hour the earth receives more energy from the sun than is consumed by mankind in a year. Unfortunately, although solar radiation is universally distributed and freely available, its low power density and intermittent nature make its collection and conversion expensive. There are a number of obstacles, both technical and non-technical, to be overcome before solar energy can become a significant part of the solution to the world's energy problems, but progress towards this end is undoubtedly being made in several respects.

The sun is of course the source of all our traditional energy resources. In addition to sunlight (or solar radiation) energy, indirect solar energy is available in the form of biomass, wind, wave and hydro power. Fossil fuels such as coal and oil are for practical purposes non-renewable forms of biomass. Only geothermal and nuclear energy are resources that do not originate in the sun.

Although mankind has put the heat and light emanating from the sun to practical use for thousands of years, it is only in recent years, particularly since the oil crisis in the early 1970's, that major efforts have been made worldwide to harness solar energy on a commercial scale. The best known approach is to convert the solar radiation into thermal energy directly applied for heating purposes. Solar water heaters are now technically and economically viable in many countries and are being installed in increasing numbers every year. More complex thermal systems are being developed for converting solar thermal energy into mechanical power for applications such as pumping and electricity generation.

Solar radiation can however be converted directly into electricity using semiconductor devices known as photovoltaic (PV) cells. Such cells had long been used in photovoltaic exposure meters, but it was not until the 1950's that higher performance devices became available that could be used for powering small specialist equipment in remote areas. The technology was greatly advanced in the 1960's when photovoltaic generators first came into regular use for space satellites. Since the oil crisis in the early 1970's, there has been a growing worldwide programme to develop and apply photovoltaic systems for terrestrial applications. Costs were initially very high, but over the last 10 years improvements in manufacturing techniques and increased volume of production have enabled photovoltaic cell costs to be greatly reduced. With further cost reductions foreseen, the possibility now exists for photovoltaics to become a major energy resource for many countries, even for those not considered to have a notably sunny climate.

Photovoltaic systems should not be thought of solely or even primarily in the context of large central power generation. Because of the distributed nature of sunlight, photovoltaics are particularly suited to small generators built close to the electrical load centres. They can work in conjunction with the main electricity grid or they can be used to provide power in remote regions far from the grid, where the alternatives are either no power at all or expensive batteries or diesel generators. Photovoltaic systems are modular and generators of any voltage and power can be built from standard modules, the overall conversion efficiency being largely independent of size. If one of the modules fails, the system may be designed to continue to operate until a replacement is fitted.

Photovoltaic systems have other advantages compared with conventional means of generating electricity. The absence of mechanical moving parts reduces maintenance requirements and makes for reliable, long lasting systems. The equipment is silent in operation, gives rise to no harmful waste products and can convert diffuse, low intensity light at an efficiency comparable to that obtaining in bright sunlight. Photovoltaic systems thus offer the possibility of a renewable energy resource with few inherent environmental disadvantages.

1.2 European Solar Energy Programme

European countries, in common with the rest of the world, have growing energy problems. Most national governments have created special departments to co-ordinate activities to find new and renewable sources of energy and, equally important, to identify energy conservation opportunities. Energy studies and research and development projects covering the full range of new and renewable sources of energy are now well established in almost every university and institute of higher education. Industry is also becoming increasingly aware of the need to introduce new energy techniques and often there is a fruitful partnership between academic research and commercial interests.

In 1975, the Commission of the European Communities (CEC) took a major initiative to stimulate and strengthen solar energy research and development activities in member countries, particularly in the fields of:

    Solar energy applications to dwellings

    Thermal-mechanical solar power plants - the helio-electric 1MWe power plant EURELIOS

    Photovoltaic power generation

    Photochemical, photoelectrochemical and photobiological conversion

    Energy from biomass

    Solar radiation data

This first solar energy R&D programme was completed in 1979 and fully achieved its objectives. The second European Communities programme, with a total budget of 46 million ECU, started in July 1979 and will run to June 1983. In the second programme, the broad headings have been maintained and two new activities added, namely:

    Wind energy

Solar energy in agriculture and industry.

There is however a significant change of emphasis in these activities: instead of the earlier exploratory work, the main emphasis is now on the development and construction of prototype systems with the aim of identifying any problem areas in systems, thereby increasing the credibility of solar energy and encouraging the rapid implementation of viable systems.

The CEC policy has been to award research and development contracts to industry, universities and research institutes, usually after publication of specific calls for tenders in its Official Journal. The contracts provide for up to 50% payment of the costs of approved proposals, the balance being raised from other sources. The funding of projects in the second CEC solar energy programme (1979-1983) is given in Table 1.1, from which it may be seen that Project C, Photovoltaic Power Generation, commands the dominant share, over one third of total budget.

Project C, Photovoltaic Power Generation, is made up of two parts: some 35 contracts for basic research and development and a further 17 contracts for photovoltaic pilot plants ranging from 30 to 300kW peak electrical output. A list of the research and development contracts is given in Appendix A and a list of the photovoltaic pilot plants is given in Appendix B.

The pilot plants cover a wide variety of applications including rural and island electrification, water pumping, ice making, power supply to a TV transmitter and hydrogen production for a factory manufacturing semi-conductors. In some instances, the pilot plants will be combined with other energy generators, such as diesel or wind generators, and many of the installations will feed any surplus electricity into the public electricity grid.

The objectives of the current CEC Photovoltaic Power Generation project are to continue research and development efforts on solar cells and arrays so as to reduce costs and increase life time, and to design and develop a family of photovoltaic power systems in the power range 30kW to 300kW set up as pilot projects for various applications covering the whole climatic range of Europe.

The CEC Photovoltaic Power Generation project is managed by Directorate General XII for Science, Research and Development, but it should be noted that photovoltaics have been supported in other ways by the CEC. Certain research and development tasks, in particular activities associated with the testing of photovoltaic cells and modules, are directly undertaken at the European Community's Joint Research Centre at Ispra, Italy. Directorate General VIII for Development is currently providing funds for some 15 projects involving photovoltaics in developing countries, as listed in Appendix C. In addition to specific applications such as pumping or power supplies, some of these projects also include technical assistance for establishing solar energy research, development and demonstration centres involving photovoltaics. This support is very much within the spirit of Art.76 of the ACP-EEC Convention of Lome (II) which provides for 'implementation of alternative energy strategies in programmes and projects that will ... cover inter alia wind, solar, geothermal and hydro-energy sources.'

PROJECT	BUDGET in million ECU
Project A Solar energy applications to dwellings	8.3
Project B Thermo-mechanical solar power plants	$4.7^{(a)}$
Project C Photovoltaic power generation	$15.9^{(b)}$
Project D Photochemical and other processes	2.6
Project E Energy from biomass	$7.4^{(c)}$
Project F Solar radiation data	2.0
Project G Wind energy	1.0
Project H Solar energy in agriculture and industry	0.7
Management and reserve	3.4
	Total 46.0

(a) includes 4.2 for Eurelios 1MWe project

(b) includes 10.0 for PV pilot plants 30 - 300kW

(c) includes 3.5 for methanol from wood, etc.

Table 1.1 Funding of projects in the second CEC solar energy
r & d programme

Within Europe itself, a number of demonstration projects involving photovoltaics have also been supported by Directorate General XVII for Energy. These are listed in Appendix D.

Finally, it should not be overlooked that in addition to providing funds for research, development and demonstration projects and for overseas aid, the CEC also takes an active role in promoting international co-operation by sponsoring workshops, meetings of contractors and international conferences. A series of international conferences of major importance in the field of photovoltaics have been sponsored in Europe – Luxembourg (1977), Berlin (1979) and Cannes (1980), with the fourth held at Stresa in May 1982. In addition an important conference on the Non-Technical Obstacles to the Use of Solar Energy was held in Brussels in 1980.

1.3 The potential for photovoltaics in Europe

In recent years, substantial investments have been made in all aspects of photovoltaic technology, from research and development through to production facilities and field installations. To date, it has been estimated that worldwide over US $1000 million has been expended on photovoltaics by public and private agencies. From the beginning, solar energy has taken the major share of the funding available for the CEC energy R&D programmes, and out of that share, the largest proportion has been spent on photovoltaics. Such a level of support reflects the high hopes placed in this technology. The question remains, can these hopes be realised? Will photovoltaics be technically and economically viable for widespread use in Europe, USA and in the developing countries within 10 years – less than the time it takes for a conventional or nuclear power plant to be planned, designed, constructed and commissioned? Could a significant proportion of Europe's total electricity production one day be derived from photovoltaics? If so, how much and when? What would be the implications of such a development?

This study addresses these important questions. After reviewing the present status of the technology of photovoltaic materials, components and systems, the prospects for a wide range of applications within Europe are then considered in detail. This is followed by a discussion of the prospects for photovoltaics in developing countries, with associated opportunities for European industry for technology transfer and for manufacturing joint ventures with local interests. The present status of the photovoltaic industry in Europe is then reviewed with particular reference to international links already established or planned.

The widespread introduction of photovoltaic power generation in Europe and elsewhere would have far-reaching social and environmental implications. These are discussed along with associated legal and institutional issues in Chapter 6.

In a concluding section, the possibilities for photovoltaics to be a potential energy resource of real significance for Europe are discussed. A possible scenario for the near, medium and long term implementation of photovoltaics in such a role is offered, with suggestions for future initiatives by the Commission of the European Communities, particularly in regard to research, development and demonstration activities.

This report concludes with a Glossary of Terms and Units, plus a number of Appendices giving details of European Community supported activity in the field of photovoltaics.

CHAPTER TWO - CURRENT STATUS OF PHOTOVOLTAIC TECHNOLOGY

2.1 Concept of photovoltaic systems

A photovoltaic generator consists of photovoltaic cells mounted in modules which form part of an array. The cells are electrically connected and the modules are in turn interconnected by wires to take the electricity to some form of control and power conditioning system. In the simplest systems, all that is required is a switch to isolate the array from the electrical load. The term 'photovoltaic system' is applied to the set of components needed to convert solar energy and supply electricity to the load, and sometimes including the load as well.

If power is required when the sun is not shining, energy storage, such as a battery, must be provided. Alternatively the system can operate in conjunction with the electricity grid or a back-up diesel generator. Blocking diodes must be incorporated to prevent electricity flowing from batteries or other generators back into the cells when they are not producing electricity. The power produced by any photovoltaic device is direct current (dc), and conversion to alternating current (ac) at standard voltage and frequency is often required, as illustrated in Figure 2.1.

For many applications, the system can be designed to 'stand alone', with no connection to the utility grid. An isolated village may thus be provided with power and a system of automatic load management can be incorporated in the control unit to turn off power to less important equipment at night or when clouds reduce the solar irradiance. This controlled load shedding ensures that power will be available where and when it is most needed without having to oversize the system.

The components of the system other than the photovoltaic modules themselves are often referred to as Balance-of-System (BOS). Whilst often involving more familiar engineering technology, it is important to develop all BOS components in appropriate ways to optimise the complete system. Considerable research and development efforts are currently being made to develop items such as cheap yet durable array support structures and reliable dc to ac inverters having high efficiency throughout their operating range.

The system designer must optimise the system and individual elements in the system to ensure that the life cycle cost is minimised. The more efficient the components and hence the overall system, the less the area of expensive photovoltaic cells that are needed and this in turn reduces the area-related costs of modules, array support structures, circuitry and land. To achieve a higher efficiency, higher costs are usually incurred and thus the technical and economic challenge is to obtain an effective balance between efficiency and costs.

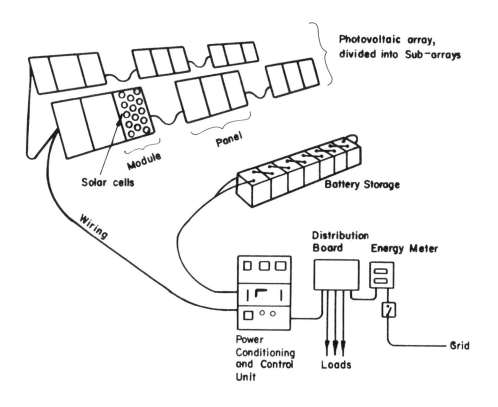

Photovoltaic array,
divided into Sub-arrays

Panel

Module

Solar cells

Battery Storage

Wiring

Distribution
Board   Energy Meter

Power
Conditioning
and Control
Unit

Loads

Grid

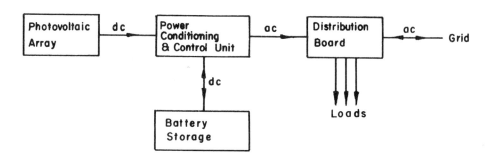

| Photovoltaic Array | dc → | Power Conditioning & Control Unit | ac → | Distribution Board | ac → | Grid |

Battery Storage — dc ↑

Loads

Fig 2.1 Typical photovoltaic system

- 8 -

2.2 Solar cells

Brief history

The so-called 'photovoltaic effect' was first observed by Becquerel in 1839 who noticed that when light was directed onto one side of a simple battery cell a voltage was developed (1). Years later, selenium and cuprous oxide photovoltaic cells were developed, leading to many applications including photographic exposure meters. It was not until the late 1950's however that solar cells with an acceptably high conversion efficiency for power generators were developed.

Silicon solar cells soon found an application in the space programme and the first solar-powered satellite, Vanguard I, was launched in 1958. Its solar generator continued to function long after the original purpose of the satellite had been achieved. Practically all satellites launched since then have been powered by silicon cells. Costs were initially extremely high but improvements in manufacturing technology and increased production volume in the late 1960's enabled unit costs to be reduced by a factor of ten.

Since the 1973 fuel crisis, interest in photovoltaics as a terrestrial source of power has increased greatly and now many countries, including several developing countries, have instituted photovoltaic research, development and demonstration programmes. These have the objective of developing a technically and economically viable power source for the full range of applications, from those needing just a few watts to multi-megawatt central power stations.

A decade ago, silicon solar cells encapsulated into modules cost over ECU 100 per peak watt of delivered power. As production techniques have improved and manufacturing volume has increased, the module price has been steadily falling and in mid-1982 was about ECU 8 to 11 per peak watt for large orders, equivalent to about ECU 6.5 to 9.5 per peak watt at 1980 prices. The objective of the world wide research and development effort is to achieve production techniques and volumes that will enable modules to be sold for less than ECU 1.00/Wp and complete systems for ECU 3.00/Wp or even less, depending on the application.

More detailed discussion of the economics of photovoltaic systems is given later in this chapter, but it should be noted that there is already an active commercial market for a number of relatively small but nonetheless important photovoltaic systems for applications such as marine navigation aids, cathodic protection systems for oil and gas pipelines, telecommunications and small pumping systems for water supply. As development continues, photovoltaic systems are expected to provide electricity cheaper than diesel and gasoline generators, the present mainstay of isolated communities. It is also expected that photovoltaic generators will be used to augment grid supplies for domestic, commercial and industrial applications and, in the long term, for central generating stations operated by utility companies.

The photovoltaic process

The interaction between photons and electrons is the basic principle underlying all photovoltaic devices. The energy generated continuously by the sun is radiated as a stream of photons of various energy levels leaving the surface of the sun in all directions. The total radiant power from the sun received by a surface of unit area is known as the 'irradiance'. Outside the earth's atmosphere, the irradiance on a plane normal to the solar beam amounts to about 1367 W/m^2, with a small variation due to the seasonal change in the distance from the earth to the sun.

As the radiation passes through the earth's atmosphere, a considerable amount is lost by scattering and absorption, some wavelengths being affected more than others. The amount of energy lost depends on the path length through the atmosphere and the amount of dust and water vapour at the time. The term 'air mass' is commonly used to denote the length of path traversed through by the atmosphere by the direct solar beam, expressed as a multiple of the path traversed to a point at sea level with the sun overhead. Air Mass 1 (AM 1) is the path length to sea level with the sun directly overhead, but Air Mass 1.5 (AM 1.5) is generally more appropriate for latitudes between 30° and 60°.

Figure 2.2 shows the spectral energy distribution of direct sunlight at sea level on a clear day for AM 1.5. The spectral energy distribution outside the atmosphere (AM 0) is also shown for comparison. This bears a marked similarity with the curve for a black-body radiator at 5900 K, the temperature of the surface of the sun.

The irradiance at ground level is made up of two components: direct and diffuse. The sum of these two components is termed the 'global irradiance'. The diffuse component can vary from about 20% of the global on a clear day to 100% in heavily overcast conditions. On a clear day in the tropics, with the sun high overhead, the global irradiance can be as high as 1000W/m^2 but in northern Europe it rarely exceeds 850W/m^2, falling to less than 100W/m^2 on a cloudy day, with most of the light in the blue/violet band. Further discussion of the solar energy reaching the earth at different times and places is given in Chapter 3.

Knowledge of the solar radiation reaching the photovoltaic cell is important for the following reasons:

- the different solar cell materials show varying levels of response to the different wavelengths of light,

- the power output from a solar cell is dependent upon the intensity of the total incoming radiation.

In order to specify or compare solar cells it is normal to quote the power output (watts peak) at an irradiance of 1000W/m^2, with standard spectral energy distribution of AM 1.5 direct sunlight and a cell temperature of 25°C (2).

The photovoltaic process, like other energy conversion processes, is subject to a maximum theoretical efficiency dependant upon the physical characteristics of the materials. The achievement of improved working efficiencies, closer to the theoretical maximum, is therefore a major

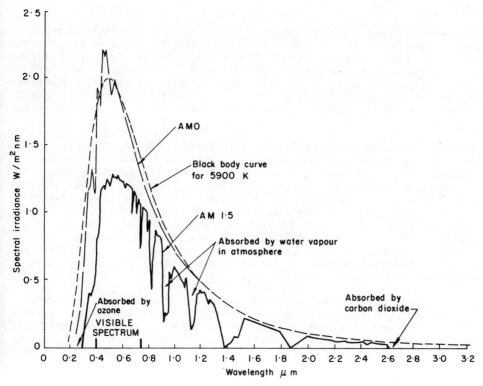

NOTE: The total irradiance under the AM1.5 curve amounts to 834.6 W/m^2

Fig 2.2 <u>Spectral energy distribution</u>

objective of research and development work. The conversion (or cell) efficiency is defined as the ratio of the maximum power output to the product of area and irradiance expressed as a percentage:-

$$\text{efficiency} = \frac{\text{maximum power}}{\text{irradiance x gross cell area}} \times 100\%$$

(It should be noted that research workers often quote the efficiencies of cell devices on the basis of active cell area, which is the gross cell area less the area occupied by the contact grid. Care is thus needed when comparing cell efficiencies).

Silicon solar cells

Detailed descriptions of the operation and performance characteristics of silicon solar cells may be found in the literature (a particularly clear account is given in reference 3) and only a brief introduction is given here.

The material most commonly used at present to make photovoltaic cells is mono-crystalline silicon. The essential features of this type of cell are shown in Figure 2.3. It is made from a thin wafer of high purity silicon crystal, doped with a minute quantity of boron. Phosphorous is diffused into the active surface of the slice at high temperature. The front electrical contact is made by a metallic grid and the back contact usually covers the whole surface. The front surface has an anti-reflective coating.

The phosphorous introduced into the silicon gives rise to an excess of what is known as conduction-band electrons and the boron an excess of valence-electron vacancies, or holes, which act like positive charges. At the junction, conduction electrons from the n (negative) region diffuse into the p (positive) region and combine with holes, thus cancelling their charges. The opposite action also occurs, with holes from the p region crossing into the n region and combining with electrons. The area around the junction is thus 'depleted' by the disappearance of electrons and holes close by. Layers of charged impurity atoms, positive in the n region and negative in the p region, are formed either side of the junction, thereby setting up a 'reverse' electric field.

When light falls on the active surface, photons with energy exceeding a certain critical level known as the bandgap or energy gap (1.1 electron Volts in the case of silicon) interact with the valence electrons and elevate them to the conduction band. This activity leaves 'holes', so the photons are said to generate 'electron-hole pairs'. These electron-hole pairs are generated throughout the thickness of the silicon in concentrations depending on the intensity and spectral distribution of the light. The electrons move throughout the crystal and the less-mobile holes also move by valence-electron substitution from atom to atom. Some recombine, neutralising their charges and the energy is converted to heat. Others reach the junction and are separated by the reverse field, the electrons being accelerated to the negative contact and the holes towards the positive. A potential difference is established across the cell which is capable of driving a current through an external load.

The generated current is built up from increments produced by photons of different energy levels (wavelengths). The high energy (short wavelength) photons are absorbed near the surface while the longer wavelength photons penetrate deeper, most being absorbed within 100μm. By plotting the incremental current generated by unit irradiance against wavelength, the 'absolute spectral response' is obtained, as shown in Figure 2.4.

By multiplying the ordinates of the absolute spectral response by the ordinates of the spectral energy distribution of the incident radiation and integrating, the total generated current I_G is obtained (Figure 2.3). The load current I_L is the difference between the generated current I_G and the junction current I_J. In the short-circuit condition (V = 0), I_J is very small. As the cell series resistance R_S is also very small for most types of solar cell, the short-circuit current provides a useful measure of the generated current. The open-circuit voltage is about 0.6V at 25° C for crystalline silicon cells.

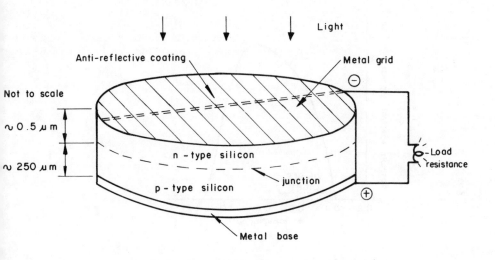

Light

Anti-reflective coating

Metal grid

$\ominus$

Not to scale

$\sim 0.5 \, \mu m$

$\sim 250 \, \mu m$

n - type silicon

junction

p - type silicon

$\oplus$

Load resistance

Metal base

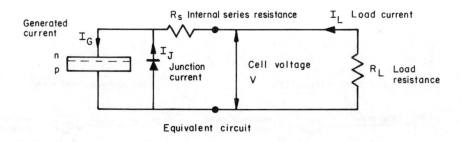

R_S Internal series resistance

I_L Load current

Generated current

I_G

n
p

I_J
Junction current

Cell voltage

V

R_L Load resistance

Equivalent circuit

Fig 2.3 Silicon cell

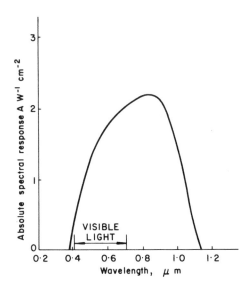

Fig 2.4   Spectral response of silicon cell

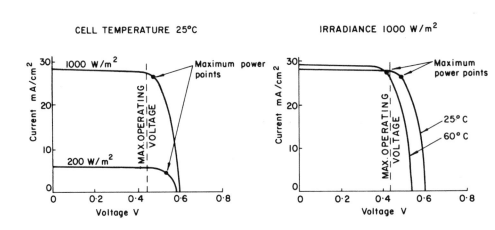

Fig 2.5(A)   I-V characteristics for
             different irradiances

Fig 2.5(B)   I-V characteristics for
             different temperatures

The current/voltage (I-V) characteristic for a typical silicon cell is dependent on irradiance and temperature, as illustrated in Figure 2.5(a) and (b). The fill factor is the ratio of the maximum power to the product of short-circuit current and open-circuit voltage. In general, the higher the fill factor, the better the cell as a practical photovoltaic device. Maximum power is represented by the area of the largest rectangle that can be fitted under the curve. The output current is practically constant for all voltages up to a voltage close to that for peak power, and is proportional to cell area. Figure 2.5(b) shows that as temperature increases the current increases slightly and the voltage decreases; in consequence the maximum power obtainable decreases. It is therefore desirable to operate the cells at as low a temperature as possible.

Silicon material

Silicon photovoltaic cells require the use of very pure silicon. The best source of silicon is silica (silicon dioxide), which occurs naturally in great abundance in the form of quartz rock and sand. Quartz rocks are reacted with a mixture of charcoal, coke and other carbon-based reductants in an arc-furnace to produce 'metallurgical grade' silicon, which cools into blocks or ingots. Many thousands of tonnes are made annually and the cost is about ECU 1.10-2.20/kg.

Metallurgical grade silicon contains about 1% impurities and these must be reduced to about 1 part in 10^9 for semi-conductor applications in electronics. This refining process is normally achieved by what is known as the Siemens process which has been used for some 20 years to provide semi-conductor grade silicon for the electronics industry. It is a high temperature, slow batch process with high specific energy consumption. The product, polysilicon granules, is consequently expensive, of the order ECU 50-70/kg. Annual world production is about 3600 tonnes and rising steadily to meet demand for semi-conductors in the electronics industry.

Until now, a small proportion of this annual production of semi-conductor grade silicon has been diverted for making photovoltaic cells. There has been some concern that the supply may not prove sufficient to meet rising demand and some photovoltaic companies are planning to secure their sources by constructing their own silicon purification plant. Fortunately photovoltaic devices do not have to use silicon quite as pure as semiconductor grade and much effort is being directed to the development of 'solar grade' silicon processes which will give an acceptable material at costs around ECU 10-15/kg. The projected substantial fall in the cost of photovoltaic cells is in part dependent on achieving this objective. Organisations active in this field include Wacker-Chemitronic (FR Germany), Semix Inc (USA), Heliosil (Italy), Dow Corning (USA), Union Carbide (USA) and Westinghouse (USA). Japanese organisations are also actively working on the development of solar grade silicon processes.

The suitably purified silicon still needs to be processed into a crystalline form suitable for the photovoltaic process. The most commonly used process to date and other promising methods being developed are described below.

Mono-crystalline silicon

The mono-crystalline silicon solar cell is an extremely stable device in the terrestrial environment and is based on well-established semi-conductor technology developed over many years for transistors and integrated circuits. Almost all cells available commercially were made this way until very recently.

Most silicon cells are 75mm or 100mm diameter, made from thin wafers cut from a long single crystal. Conversion efficiences currently range from 12 to 18%, the theoretical limit being about 23%. Rather than attempting to improve the efficiency of commercial silicon solar cells, the present trend is to find the optimum balance between efficiency and the use of lower cost solar grade material. The long single crystals are usually made by what is known as the Czochralski (Cz) process, whereby an ingot is drawn out of a melt of pure silicon. The atoms of silicon solidify in a perfect cubic pattern following the structure of a seed crystal immersed in the melt and slowly drawn up at a speed of up to 100mm/hour. The process is time and energy consuming and the resulting crystal ingots are expensive, costing over ECU 100/kg. Although interesting attempts are being made to develop advanced Czochralski techniques, involving continuous production of ingots (4), the use of the Czochralski method may decline with the advent of new polycrystalline silicon cell techniques.

Cast silicon ingot processes

Several groups are developing cast ingot processes and in a few cases the method has reached the stage of commercial production. One such is the heat exchanger method (HEM) being developed by Crystal Systems Inc (USA) (5). Directional solidification of silicon ingots is achieved from a melt, the temperature being controlled by a heat exchanger, with helium gas as the cooling medium. The ingot produced is over 95% single crystal and cell efficiencies up to 15% have been achieved for 300 x 300 x 150 mm ingots.

Polycrystalline silicon ingots are being commercially produced by two other companies: Semix Inc (USA), a company associated with Solarex Corporation (USA), who describe their product as semi-crystalline silicon, and Wacker-Chemitronic (FR Germany), who call their product 'Silso'. The reported efficiency of commercial cells made by these two processes is at least 9% and the cost of production is considerably less than for the Cz method.

The Semix material is being produced by a proprietary process developed under a joint agreement with the US Department of Energy (6). Solar grade silicon can be used in this process, giving further important cost reductions. Solarex are expecting that Semix technology incorporated in their new range of modules will rapidly replace their current monocrystalline range.

The Wacker-Chemitronic Silso material has been developed with the support of the government of the Federal Republic of Germany (7). They also hope to be able to reduce costs in a few years by the introduction of solar grade silicon. Figure 2.6 shows a Silso wafer before it is made into a cell - the polycrystalline structure can be clearly seen.

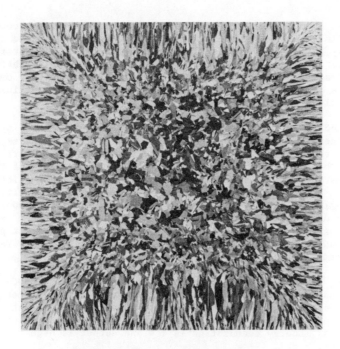

Fig 2.6   A wafer of Silso polycrystalline silicon
          - original size 100mm x 100mm
          (Source: Wacker Chemitronic)

Another cast silicon ingot process is being developed in France by the CGE
Laboratoires de Marcoussis, with support from the French government and the
CEC (8).   In this process, polycrystalline ingots are formed by the
orientated solidification of molten silicon in a crucible under directional
heat flow.   Industrial production is planned to start in 1982.

All ingot processes, whether mono or polycrystalline silicon, have the
drawback that they involve sawing to form wafers at least 350 µm and often
as much as 400–450 µm thick.   Sawing is expensive   because   it   is
timeconsuming and wasteful — over half the material is lost as kerf (ie
sawdust). Efforts are being directed towards the improvement of sawing
techniques, to give a high yield of thinner wafers, about 250 µm thick, by
such means as saws with thinner blades, multi-bladed circular saws and wire
saws, some cutting as many as 1000 wafers at a time.

Continuous ribbon processes

There are several continuous ribbon processes under development, all
seeking to demonstrate the potential for achieving low cost, high
efficiency solar cells.   The most promising methods are described below.

The dendritic web growth method produces a thin, wide ribbon of single
crystal silicon directly from a melt of molten silicon.   'Dendritic' refers

to two wirelike dendrites on each side of the ribbon and 'web' refers to the liquid film supported by the bounding dendrites. The process produces high efficiency cells and there is little or no waste of material as cutting is not required apart from removal of the dendrites. Westinghouse Electric Corporation (USA) are developing this process, with the objective of achieving a throughput rate of 2500mm^2/minute and a cell efficiency of 15% (9). A small pilot production plant is now operating.

The edge-defined film-fed growth (EFG) method is being developed by Mobil Tyco Solar Energy Corporation (USA) (10). The process involves feeding molten silicon through a slotted die, whereby the shape of the ribbon is determined by the contact of molten silicon with the outer edge of the die. The die is constructed from material that is wetted by molten silicon, such as graphite. Efforts are being directed to extending the capacity of the EFG process to increase the speed of production and width of ribbon. Mobil Tyco have also demonstrated large octagonal cylinders of ribbon silicon, thereby avoiding edge problems. Cells of 13% efficiency have been produced from EFG silicon ribbon. Mobil Tyco say that this process is now ready for large-scale commercial production.

Honeywell Inc (USA) is developing the silicon-on-ceramic (SOC) process which continuously creates a thin film of polycrystalline silicon on an inert substrate (11). Cell efficiencies are as yet below 10% but the thin film requires only one quarter of the silicon used in methods which involve sawing of ingots. Energy consumption is low and output is high in relation to the equipment cost.

The ribbon-against-drop (RAD) process is being developed in France by LEP (12). In this process, a thin layer of molten silicon is deposited on a graphite-coated Papyrex ribbon as it is moved upwards through a crucible. The production rate is being steadily improved but the process is still some way from commercial introduction.

Before proceeding to discuss thin film cells, it is worth noting that all the main crystalline silicon techniques described above are said by their proponents to be capable of achieving cost targets for complete modules of less than ECU 1.00/Wp, given large enough production volumes and reasonable success with the development of cheaper materials and production processes. It has also been reported that even the current Cz and polycrystalline silicon technology is capable of achieving ECU 1.00/Wp from plants producing annually sufficient modules to provide 100 MWp of power (13). This is also the conclusion to be drawn from estimates prepared for the Laboratoires de Marcoussis polycrystalline silicon process which indicate that it should be possible for a large plant to make wafers for as low as FF 1.35/Wp (ECU 0.20/Wp) based on the assumption that solar grade silicon will in time become available at FF 60/kg (ECU 9.1/kg) and that low cost wafering techniques are developed (8). With wafers available at this cost, complete modules should then be possible for about FF 4/Wp (ECU 0.61/Wp).

Further discussion on the scope for cost reduction for photovoltaic modules and complete systems is given at the end of this chapter.

Thin film solar cells

Many thin film materials and technologies are undergoing development and evaluation for potential commercialization. Because the photosensitive layers are very thin, often much less than 5μm, very little material is used and costs are low. Present research efforts are being directed to improving efficiency and long-term stability with reproducible parameters. The processes are of varying complexity and have various potentials for cost reductions. The two most promising lines of development are amorphous silicon and cadmium sulphide/cuprous sulphide cells, together with related compounds.

Many groups are working to develop amorphous silicon (a-silicon) cells. An efficiency of about 15% is possible in theory and it is generally considered that an efficiency of at least 8% is required for large-scale commercial applications. Pocket calculators and watches which incorporate small a-silicon cells that are particularly sensitive to fluorescent light are being produced in great numbers by Sanyo (Japan), but there are difficulties in progressing to large area devices. As with other thin-film cells, high series resistance with resulting low fill factor is a serious problem. Pioneer work on low cost hydrogenated a-silicon cells was carried out at the University of Dundee (UK) and major development efforts are continuing there, as well as in Japan and in USA. RCA (USA) are confident that they can achieve 8% efficient cells (14) and Energy Conversion Devices (USA) maintain that module costs less than US $0.50/Wp (ECU 0.49/Wp) will be possible with mass production of their fluorine-silicon cells (15).

Practically the entire Japanese photovoltaic research and development effort has been concentrated on the development of amorphous silicon cells, which they see as the most promising route to low cost photovoltaics. A number of commercial companies and universities are involved in an integrated programme sponsored by the Ministry of International Trade and Industry, the objective being to establish a major industry making amorphous silicon cells with efficiency at least 8%, with module costs US $ 0.50/Wp (ECU 0.49/Wp) or less. Hamakawa reports that significant progress was made in 1981 (16). The best efficiency for a-silicon cells reported to date is 8.04% by the research team at Osaka University for small area cells. The efficiencies for large area cells suitable for practical applications have also been improved to between 5.6 to 7.1% by Fuji, Sanyo, TIT and Osaka University. Sanyo are building up invaluable experience in the commercial production of amorphous silicon cells which are being used in large numbers to power watches, calculators and portable radios and expect to improve both efficiency and cell area by decreasing the reflectivity, optimising the p-i-n structure and increasing the film uniformity.

The other contender for low cost photovoltaics is the cadmium sulphide/cuprous sulphide cell, or one of its variants. Very thin cells are produced by continuous deposition of various layers onto a substrate. A number of groups have been developing such cells for many years but until recently efficiencies have been too low and lifetimes too uncertain for significant commercial sales, despite major efforts in the late 1970's by the pioneering company Solar Energy Systems (SES) of the USA.

Photon Power (USA) has however completed a commercial-scale manufacturing facility and expect soon to be producing large area cells with efficiency at least 4% (17). They expect in the course of time to be able to improve on this efficiency, thereby effectively increasing the plant capacity from its present 5 MW/year to as much as 10 MW/year. Nukem (F R Germany) is also constructing a cadmium sulphide cell plant based on a sputtering process originally developed at the University of Stuttgart and expect to be in commercial production at the end of 1982. In the UK, Thorn-EMI is developing a cadmium sulphide/cuprous sulphide process based on electrophoretic deposition of the thin films (18). Although encouraging progress is being made, it is likely to be several years before the process is ready for commercial production, although module costs of less than ECU 1.00/Wp appear to be readily achievable. The University of Delaware (USA), using a more sophisticated process, have demonstrated 10% efficiency for small cadmium sulphide/cupric sulphide cells in the laboratory (19). It is not yet clear whether large area cells could be produced by this method on a continuous, low cost basis, although their studies indicate that module costs as low as US $ 0.30/Wp (ECU 0.30/Wp) may be possible (20).

The imminent commercialisation of new and improved amorphous silicon and cadmium sulphide cell processes will undoubtedly provide a strong challenge to the present dominance of crystalline silicon technology in the market.

Gallium arsenide cells

Gallium arsenide is an interesting material for photovoltaic applications for two reasons. First it is more efficient as an energy converter than other materials, and, second, it does not lose efficiency at high temperatures as rapidly as silicon. Gallium arsenide is thus particularly suitable for use in high concentration modules which focus the incoming direct radiation onto a small area of active cell. Efficiencies approaching 25% have been demonstrated with single crystal gallium arsenide cells and 17% has been obtained with thin film gallium arsenide devices (21). Production costs however are high and gallium arsenide is not expected to be a significant competitor for very low cost photovoltaic systems.

Advanced cell types

There is much research going on worldwide into advanced photovoltaic devices which may eventually become attractive commercial propositions. Among the more interesting may be mentioned the following:

- Texas Instruments (USA) is developing a solar cell-plus-electrolyte system, using small spherical silicon collectors with either n centres and p coatings or p centres and n coatings (22). The process is shown schematically in Figure 2.7. Metal electrodes form anodes and cathodes which force a current through hydrobromic acid electrolyte, separating it into hydrogen gas and liquid bromine which are separately stored and recombined in a fuel cell to generate electricity on demand as well as reconstituted electrolyte. In this closed-loop system, the collector fluid is also warmed by the sun and has to be cooled by being passed through a heat exchanger. The thermal energy thus extracted may be put to good use, such as for water heating. Considerable work is still required to develop a practical system but the potential advantages are

many, including competitive efficiency (about 13% anticipated), low
production costs and, in particular, no requirement for batteries.

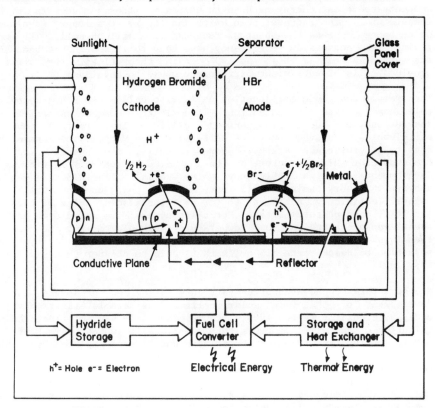

Fig 2.7   Solar-cell-plus-electrolyte system being developed by
          Texas Instruments

- Several groups in Europe and USA have been investigating split spectrum
  cells, in which the incoming concentrated light is split by filters and
  mirrors into spectral bands which are then individually directed to cells
  which are matched to the photon energy in each band. Varian Associates
  (USA) have reported efficiencies as high as 27% for a split spectrum cell
  pair of this type, with an efficiency of 20.5% for the complete
  module (23).   It is however difficult to foresee sufficient cost
  reductions to make this approach commercially viable.

- Multiple junction tandem cell systems employ three or more cells of
  different materials stacked together. The top cell absorbs and converts
  high energy photons. The next cell absorbs and converts photons with
  less energy. The next lower cell absorbs photons with lesser energy and
  so on. A total theoretical efficiency of 42.8% can be obtained for a
  seven semiconductor system (24). Although complex, such systems may in
  time become practicable given suitable low cost thin films.

- Another high efficiency approach is the thermophotovoltaic (TPV) cell being developed at Stanford University (USA). In this process, concentrated solar energy is used to heat a radiator surface to a high temperature. The radiator then emits the energy at a longer wavelength, which better matches the spectral response of a silicon cell. TPV cells giving 28% conversion efficiency have been built and it has been shown that efficiencies as high as 40% could theoretically be achieved, provided cells of sufficient performance could be developed (25).

- The Metal Semiconductor Junction cell, also called the Schottky barrier cell, has received much attention. It could in theory be as efficient as conventional p-n or n-p junction cells. In practice however, efficiencies achieved on these cells have been very low because of high density of interface states. More recently, the Metal Insulator Semiconductor (MIS) structure has been studied extensively in view of the development of high efficiency solar cells. The MIS cell can also achieve in theory the same efficiency as the p-n junction cell. The highest efficiency achieved so far on silicon solar cells of this type was 16% (based on total area). For silicon, the MIS structure has the advantage of being a low temperature technology, but to date the techniques have proved to be rather complex. It has not yet been possible to demonstrate MIS cells that could be cheaper than similar n-p type cells.

- MIS structures do however offer very interesting new prospects for the development of thin-film solar cells. Considerable efforts have been made, particularly in the 1960's, to develop Cadmium Sulphide, Cadmium Telluride and similar types of solar cells which were of the heterojunction type in which copper compounds were employed as the p material. The copper compounds proved to be relatively inefficient and gave rise to large stability problems. It now seems possible that these problems can be overcome with thin film solar cells, a good example being the Cadmium Selenide cell with an MIS structure instead of the copper compounds. The Battelle Institute, Frankfurt (F R Germany) is carrying out development work on such cells under a CEC contract and has reported 6% efficiency coupled with high stability.

Solar cell technology status

Solar cell technology is dynamic and currently undergoing many changes. Lower cost solar grade silicon material will soon be available to replace the expensive semiconductor grade material. The previous dominance of Cz monocrystalline silicon is being challenged by polycrystalline silicon and other cast silicon materials. Advanced Cz monocrystalline silicon techniques and one or more of the silicon ribbon processes may well be commercially successful when introduced on a large scale. It is important to note that at least one major company is planning soon to introduce large area cadmium sulphide cells at a price that is said will be competitive with silicon cells. It will take several years however for confidence in the long term performance of such cells to be established.

Amorphous silicon cells may also provide a low-cost competitor within a few years. The long experience and known characteristics of crystalline silicon will however ensure that this material holds on to its present pre-eminent position for at least 5 years and probably longer.

2.3 Photovoltaic modules and arrays

General description

Solar cells can be interconnected in series and in parallel to achieve the desired operating voltage and current. The basic building block of a flat-plate solar array is the module in which the interconnected cells are encapsulated behind a transparent window to protect the cells from mechanical damage and the weather. One or more modules are then attached to a supporting structure to form a panel and a number of panels makes up a solar photovoltaic array, divided possibly into a number of sub-arrays. (Note: some manufacturers use the term panel to refer to a module).

A typical module is illustrated in Figure 2.8. It consists of 36 series-connected 100mm diameter mono-crystalline silicon cells and is rated to produce 34.6Wp at 16.1V when the cell temperature is 25° C in sunlight of $1000W/m^2$ irradiance AM 1.5 spectrum. The overall dimensions are 1070mm long x 410mm wide x 38mm thick, with a weight of 5.0kg. The cells are encapsulated in a transparent resin between a window of low-iron content tempered glass with an anti-reflective surface and a backing of plastic coated aluminium sheet. The edges have a rubber sealing gasket held in place by anodised aluminium framing.

Figure 2.9 shows the photovoltaic array providing power for a large irrigation pumping installation at Montpellier, France. It consists of 768 modules arranged in 192 panels of 4. The array has a rated power of 25kWp.

Encapsulation

Photovoltaic systems must be designed for long-life with minimum maintenance. Modules are required to have a lifetime of at least 20 years and good examples will probably last much longer. In general, the structure of a flat-plate module is made up of six functional layers, although not all layers need be present for a given design (Figure 2.10):

1. Top surface layer
    Needs to be easily cleaned, abrasion resistant, anti-reflective and, as far as practicable, self-cleansing by the action of wind and rain.

2. Top cover or window layer
    Needs to be strong with good impact resistance, high transmissivity in the waveband 350 to 1200nm and UV resistant.

3. Pottant or encapsulant
    Must be flexible to permit differential movement due to differential temperature-induced expansion and bending forces, highly resistant to the ingress of water or air that could degrade the cells or the electrical connections, chemically stable, compatible with materials with which it is in contact, temperature stable, UV resistant, a good electrical insulator, and have good thermal conductivity and high transmissivity.

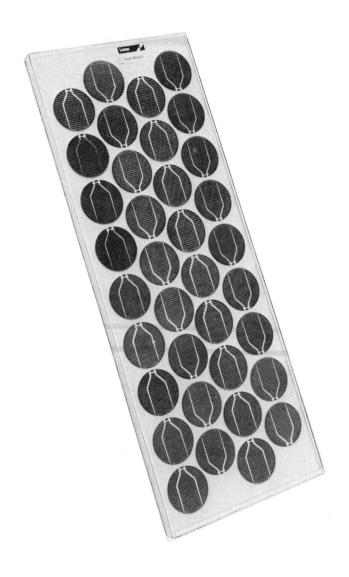

Fig 2.8 Typical photovoltaic module
         (Source: Lucas BP Solar Systems)

Fig 2.9 26kWp array for large pumping
installation at Montpellier, France
(Source: France Photon)

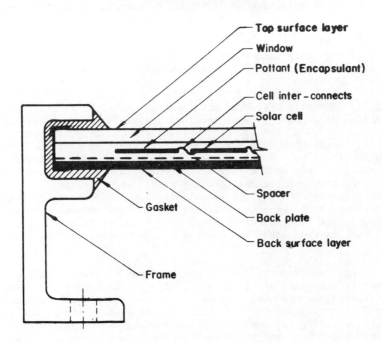

- Top surface layer
- Window
- Pottant (Encapsulant)
- Cell inter-connects
- Solar cell
- Gasket
- Spacer
- Back plate
- Back surface layer
- Frame

Fig 2.10 Construction of module

4. Spacer
   This provides electrical insulation and mechanical separation between the cell and the back plate.

5. Back plate
   Needs to be light and strong to provide structural support (may be omitted if the window layer is designed as the main structural support).

6. Back surface layer.
   Must be weather resistant.

Power output from a photovoltaic cell decreases with increasing temperature. Therefore, to keep operating temperatures as low as possible, the module should have good passive design features to enhance cooling by radiation, conduction and convection. The Normal Operating Cell Temperature (NOCT) is an important consideration for design purposes and is defined as the working temperature of the cells in the module when the irradiance is 800W/m^2 , the ambient air temperature is 20° C and the average wind velocity is 1m/s. Repeated thermal cycling calls for good design and materials to reduce the incidence of cracked cells and broken interconnects. The interconnects must allow for small relative movement between cells.

Cells must be well insulated against the high voltages possible when several modules are connected in series. Even small leaks to earth can result in significant power loss and can lead to module failure.

The module must also have adequate strength to stand up to normal handling during transportation to the site and installation. In service stresses may be induced by misalignment of the panel structure, wind loads and loads imposed by snow and ice. It should be easy to instal and connect up. Terminal boxes or cable connectors must be fully protected against ingress of water, and designed for ease of installation as well as safety and durability.

A potentially damaging reverse current can develop in the electrical circuits if the difference between the load voltage and the open-circuit voltage of one of a number of parallel strings of photovoltaic cells becomes sufficiently high as a result of a cell fault or partial shadowing. A damaging heating ('hot spot') can then arise in the reverse biased cells if the heat generated by the reverse current cannot be dissipated quickly enough by conduction and radiation. By-pass and blocking diodes are usually incorporated into the electrical circuits, to protect the cells and modules against such reverse currents that might otherwise arise under normal operating or fault conditions. It is also prudent to provide a degree of circuit redundancy to reduce the consequences of individual cell failures.

All these design requirements have of course to be achieved at minimum cost and each manufacturer has a slightly different solution. An important study of this subject has been carried out by the Jet Propulsion Laboratory (JPL) in California, USA, to 'define, develop, demonstrate and qualify encapsulation systems, materials and processes that meet the low cost solar array project life, cost and performance goals.' This study has identified a number of materials and methods that had not previously been commercially used and recommendations have been made for cost-effective encapsulation systems (13).

Tempered, low-iron glass is at present the most common window material. In many respects it is ideal but it is rather heavy and not particularly cheap. Plastics, although lighter and sometimes less expensive, are generally less stable and are not so abrasion and soiling resistant. Some types of plastic are permeable to water vapour and others degrade under prolonged exposure to UV light.

The most common pottant used to be silicone rubber, which was highly transparent and flexible, but many manufacturers now use PVB (poly-vinyl butyral), the plastic used in laminated glass, since it is cheaper and bonds better to the window and backing materials. PVB is however hygroscopic, calling for careful water sealing at the edges. Some manufacturers have opted instead for EVA (ethylene vinyl acetate), which has similar properties and is somewhat cheaper.

Glass, aluminium, steel and glass fibre reinforced plastic (GRP) have all been used as backing materials, with varying advantages and disadvantages. Some manufacturers avoid introducing a backing plate altogether, using the top glass window as the main structural support. A plastic-protected twin metal foil at the back then serves as a thermal conducting layer and weather seal. The upper surface of the backing plate should be white, to reflect light and reduce heat gain.

There have been a number of developments recently to introduce semi-automated production methods for module construction. Many operations however, such as frame assembly and the fitting of terminal boxes, are presently labour intensive and time-consuming. Attempts are continuing to develop continuous production methods using laminated plastics and resins but the practical problems are formidable.

Cell and module testing

As any new technology develops it is normal for appropriate testing methods and ways of specifying the qualities of the product to be developed in parallel. Test procedures are first established for the proper definition of the performance and characteristics of the new device. Later, attention turns to considerations of quality control in mass production, the performance requirements of the end-users and the development of accepted standards in the industry for the products.

Throughout the extensive programme of photovoltaic development work sponsored by the US Department of Energy since 1975 attention has been given to the development of appropriate test procedures. The Jet Propulsion Laboratory (JPL) has been responsible for the establishment of cell and module test procedures to qualify equipment offered for inclusion in the various block purchases of the US DOE. The requirements have been revised in the light of experience and the current specification, referred to as Block V, provides for a comprehensive range of electrical performance measurements and physical durability tests.

The Commission of the European Communities has also been developing standard test procedures, based on the advice of an international committee established in consultation with European Community member countries. Specification No.101, Standard Procedures for Terrestrial Photovoltaic Performance Measurements, was issued in 1980 and this provides standards for determining the electrical performance of photovoltaic solar cells, modules and arrays. Measuring methods and instrumentation are described for testing in natural sunlight and with a solar simulator. To minimize discrepancies caused by variations in the spectral energy distribution of the incident radiation, the specification requires performance ratings to be related to a 'standard sunlight' distribution (AM 1.5) and for irradiances to be measured by specially-calibrated reference solar cells.

A companion document, known as Specification No. 501, Photovoltaic Module Control Test Specifications, was issued in August 1981. This document lays down test specifications for photovoltaic module tests of the Commission of the European Communities which are to be applied by the Joint Research Centre at Ispra, Italy, for the acceptance of prototype and production modules for pilot and demonstration projects of the Commission. It contains the test schedule and a detailed description of 20 control tests, the purpose being to provide data on the performance rating of photovoltaic modules and to identify environmental factors and design features which could affect their durability. It is hoped that this document will become the basis for manufacturing standards that will be accepted throughout the industry.

Concentrator designs

The photovoltaic modules described so far are flat plate devices that are usually mounted in a fixed plane. Another approach uses optical concentration systems of lenses or mirrors to focus direct sunlight on high efficiency silicon or gallium arsenide cells. Three main types of concentrator design have been developed (Figure 2.11), as follows:

(a) line focus parabolic trough;

(b) linear fresnel lens; and

(c) point focus fresnel lens.

These concentrators are made of plastic, glass and aluminimum and reduce the requirements for expensive photovoltaic cells. Silicon solar cells are limited to approximately 200x concentration but gallium arsenide cells can be used effectively up to concentrations of 1000x at an estimated 25% efficiency. Relatively low concentration (2-3x) can be achieved more simply by means of plane mirrors set at appropriate angles on either side of a flat plate module, to form a V-trough. Comparative performance evaluation of various concentrator options was carried out by the Joint Research Centre, Ispra, in 1978 (27).

The immediate attraction of concentrator systems is the reduced requirement of the expensive photovoltaic cells compared with flat plate modules, for the same electrical output. This advantage will become progressively less significant as the price of flat plate modules continues to fall. The cells in concentrator systems get hot, which means that special cooling arrangements are necessary. At low concentrations, passive cooling fins

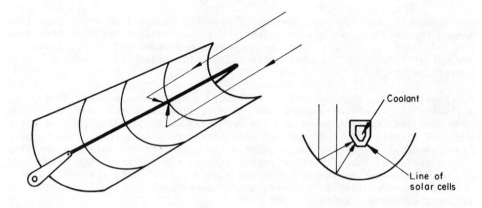

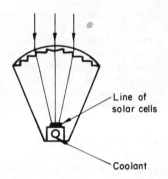

Coolant

Line of
solar cells

(a) Line focus parabolic trough

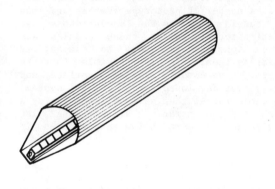

Line of
solar cells

Coolant

(b) Linear focus fresnel lens

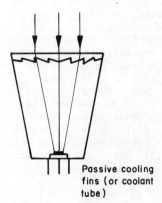

Passive cooling
fins (or coolant
tube)

(c) Point focus fresnel lens

Fig 2.11 <u>Concentrator options</u>

are adequate but active cooling by circulating water or other liquid is needed for high concentration systems. The heat removed can sometimes be put to good use, for heating domestic hot water for example, and the overall efficiency of the system effectively increased, in effect, a 'total energy system'.

Except for very low concentration systems, concentrator modules can only utilise the direct component of the solar radiation and to do this they must continually track the sun. The mechanism to do so adds to the cost and complexity of the system and introduces additional maintenance requirements. The higher the concentration, the more accurate the tracking system needs to be. In electrical terms, the overall efficiency of concentrator systems is little better than for flat plate systems and thus the land area required is similar, taking into account space needed for tracking mechanisms and controls.

The relative complexity of concentrator systems indicates that they will take longer to develop than flat plate arrays. So far much less investment has been made in their development. A number of prototype systems are now operating or under construction but in all cases the concentrators have been purpose-designed. The fabrication techniques for these modules are not yet developed for mass production, and as yet, no company is commercialising complete concentrator systems. It is possible for the focussing lens to act as the protective cover to the cell surface, and other economies in construction may yet be developed for practical systems by considering the module as a whole rather than as a cell set inside a concentrator. The testing procedures for concentrating systems are also less well developed and some difficulties are being experienced in predicting the performance of modules.

Uneven illumination of the solar cells, damage from thermal cycling, poor thermal contact between cells and substrate, deterioration of reflectors and lenses are some of the problems that have been encountered in prototype systems. The effect of wind-blown dust and condensation is more critical to concentrator systems than to flat plate systems and may pose major maintenance problem.

Estimates for mass production have been published indicating that concentrator modules could possibly be produced for less than ECU 2.80/Wp and thus be competitive with the medium-term targets for flat plate collectors (28). Given the relative complexity of concentrator systems, it may be expected that engineering costs will need to be spread over very long production runs before the module costs can be reduced to a competitive level. The target figures may therefore be more difficult to achieve than the targets for flat plate modules. If they can be demonstrated as practical propositions for large scale generation, then no doubt the necessary scale of production could be achievable. The need for direct sunlight will restrict the geographic locations at which concentrator systems can be effectively deployed and it is unlikely that they will ever be competitive in Europe, where diffuse radiation is a significant proportion of the total solar energy received, as discussed in Chapter 3.

Photovoltaic/Thermal (PVT) modules

As mentioned above, concentrator systems can also be used to provide thermal energy. Some development work is being devoted to combined photovoltaic/thermal (PVT) flat plate collectors, which similarly produce electrical power and hot water, but without involving the need for complex tracking mechanisms. Reliable systems have yet to be demonstrated and opinions differ regarding their economic viability (29). Much depends on the value placed on the hot water produced, but they do have the advantage that when space is limited, for example on the roofs of houses, the use of PVT collectors could meet both electrical and thermal energy needs whereas there may be insufficient area available for separate systems.

Array support structures

Although not involving any new technology, it is important that array support structures are appropriately designed for the lowest cost consistent with long life with minimum maintenance. Until the present, modules costs have been high and consequently there has been little incentive for designers to develop low cost solutions for support structures. As the cost of modules continues to fall, it will become increasingly important to employ low-cost materials that minimise life-cycle costs. Studies by JPL indicate that it may be possible to achieve an installed cost of mass produced structures in the range ECU 23 to 50 per square meter of module area, depending on the location and array size (13). This will represent a significant proportion, up to 50%, of the total installed cost of a future array, assuming the price of the photovoltaic modules reduces in time to less than ECU 1.00/Wp.

In some countries, treated timber is readily available as a construction material. Elsewhere, bamboo might be appropriate or light steel sections suitably protected. One highly adaptable design developed by Photowatt International using extruded aluminimum sections is shown in Figure 2.12. The cost would be prohibitive for small quantities, but by standardising on this design for all applications, Photowatt International report that they have achieved less than ECU 50/m^2.

For many applications, particularly for residential systems, the photovoltaic array may be mounted on the roof of a building. Several approaches to fixing the modules are being investigated, including some schemes in which the modules themselves form part of the roof structure. Provided adequate waterproofing is achieved and provision is made for changing defective modules if necessary, this integrated approach is likely to show significant cost advantages compared with arrays mounted on a conventional roof structure.

All array designs must take full account of the appropriate electrical and building codes. Appropriate lightning protection must be provided and all exposed metal parts must be suitably earthed. Due consideration must also be given to the environmental impact of large arrays. The sympathetic treatment of a site using well-designed structures and careful location of the array can minimise the visual impact. This is one of the objectives of the CEC sponsored photovoltaic pilot plants listed in Appendix B, including for example the 50kWp system to be built on Terschelling Island, The Netherlands, Figure 2.13.

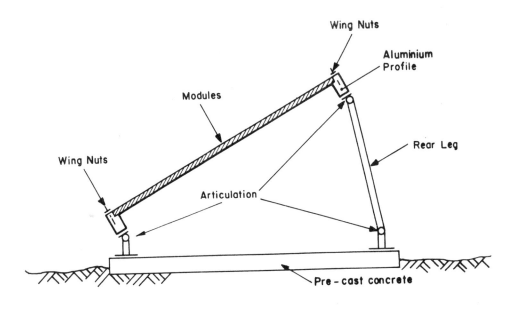

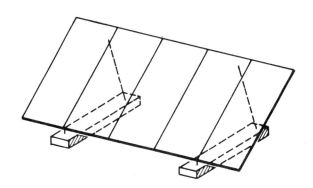

Fig 2.12 <u>Low cost array support structure proposed by</u>
         <u>Photowatt International</u>

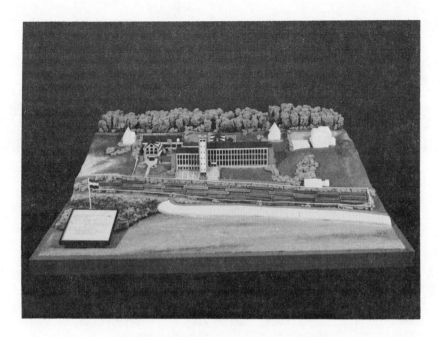

Fig 2.13 Model of 50kWp photovoltaic system
         to be built on Terschelling Island Holland
         (Source: CEC)

The fixed flat plate array is the most common type at present but there are
other options that may provide cost-effective solutions in certain
circumstances. These include the following:

1. Seasonally-adjusted tilt
   The inclination of the array is manually adjusted at intervals through
   the year to allow for the changing elevation of the sun. This is a
   simple and inexpensive way to augment the electrical energy output, but
   would be impractical for large arrays involving hundreds of panels.

2. Single-axis tracking
   The array is mounted on a structure that maintains its inclination with
   the horizontal constant but rotates about a vertical or, better, a
   polar axis to follow the sun, using a clock-regulated drive mechanism.

3. Two-axis tracking
   The array is mounted on a structure that rotates to maintain the array
   surface normal to the direct solar beam. Such systems are unlikely to
   be justified for plane arrays.

For small arrays, consisting of only a few modules such as would be
required for a small-scale pumping system, it may be convenient to provide
a portable array structure, which can be turned to face the sun at various
times in the day to maximise output. The array may in this case be moved
from place to place as required and also taken indoors at night for safe

keeping.

Overall array design

There are many factors to be considered in the design of a photovoltaic
array and it is necessary to consider the technical and economic
implications of various possible solutions before arriving at the final
design. The overall objective must be to optimise the system as a whole.
Among the factors to be included in such a study are:

i) Total power required, taking into account wiring diodes and mismatch
losses, plus any long term degradation of performance.

ii) Choice of operating dc voltage, taking into account such matters as
resistance losses in wiring, performance of power conditioning
equipment, safety and costs.

iii) Degree of reliability required, to be achieved by division of array
into sub-arrays and smaller units, with possibly parallel wiring to
reduce the impact of a module failure.

iv) Array support structure design, including orientation, inclination,
provision for adjustment, materials of construction, foundation
conditions, maintenance, module servicing, etc.

v) Diagnostic and test facilities, for individual modules, panels,
sub-arrays and the array as a whole.

vi) Provisions for earthing and lightning protection.

Mismatch losses arise from the fact that the current voltage
characteristics of individual photovoltaic modules vary. The peak power
can be as much as ± 10% from the nominal value specified by the
manufacturer. When the modules are linked in series of strings, the
combined current-voltage characteristic from one string will in general
differ from the others to which it is linked in parallel, so that certain
modules or strings are not operated at the optimal point on the
current-voltage characteristic. This is termed mismatch loss. Mismatch
losses may be reduced by selection and matching of the modules incorporated
into each string and to this end some manufacturers grade each module as a
routine part of the manufacturing process. In general however matching of
modules to reduce mismatch losses takes time and the cost-effectiveness has
to be checked for the particular circumstances.

As previously mentioned, most photovoltaic modules incorporate protection
in the form of diodes to prevent reverse currents which may result in
damaging hot spots. Strings of modules in series may also need to be
protected by diodes to prevent reverse currents, which can arise either
from a failure in some part of the array or from partial shadowing of the
array, which results in a gross albeit temporary mismatch of the
current-voltage characteristics.

For the European Community photovoltaic pilot plants, the requirements
specify that the power loss arising from wiring, diodes and mismatch losses
must not exceed 5%, as an inducement to achieve energy efficient designs.

A wide variety of array configurations have been proposed, with varying dc voltages and levels of reliability, but all the designs comply with the 5% loss limit.

Overall array design for large photovoltaic systems is a complex subject and much development work is needed to identify the most appropriate solutions for various applications. Failure analysis and reliable estimation of long-term performance are also subjects receiving increasing attention, as these are vital to the economics and hence commercial viability of photovoltaic systems.

As an example of how these various factors have been taken into account in the design of a large photovoltaic array, reference may be made to the 100kWp system designed by Siemens (F R Germany) for Kythnos Island, Greece (one of the European Community photovoltaic pilot projects). The electrical layout is illustrated schematically in Figure 2.14. The array power is defined as 100kWp, made up of four 25kWp sub-arrays. Each sub-array will contain 200 modules arranged in 10 parallel strings. Each module has a nominal power rating of 125kWp, but the contractor proposes to grade and then select modules for each string to reduce mismatch losses. After allowing for losses due to unavoidable mismatch wiring and diodes the net peak power available from the array is estimated to be about 95kWp. By providing three inverters each rated at 50kW, the plant will be able to supply a peak power of 150kW for relatively limited periods, by drawing from the batteries. The nominal operating battery voltage needs to be as high as possible to reduce resistance losses, but the upper limit is effectively determined by the power conditioning units. To achieve the required high efficiency, Siemens propose to use transistor technology for the power switching circuits in the inverters. The maximum voltage that transistors can sustain is 500V and, after allowing a safety margin, the nominal operating voltage thus has to be limited to 250V dc, with a range from 200 to 300V depending on the charge state of the battery.

Siemens have decided to interpose dc/dc converters between the four photovoltaic sub-arrays and the batteries. The superimposed control system will control these converters to maximise the power transfer (ie, they serve as maximum power point trackers). With the converter design selected, the optimum ratio of the array voltage to battery voltage is about 0.6, which explains the choice of 160V as the nominal array voltage. The array has been divided into 25kWp sub-arrays to provide a degree of reliability (only 25% loss of output in the event of a sub-array failure). In addition, Siemens consider 25kWp to be a convenient module size with which to build up larger plants in the future. This pilot plant will enable them to gain valuable experience in this respect.

Each 25kWp sub-array is made up of 10 strings of 20 modules, with the strings connected in pairs. This arrangement of wiring and terminal blocks will facilitate the location of faults. The supervisory control system compares the power output from each double string and also from each 25kWp sub-array. If a power difference exceeding a specified amount is noticed, an alarm is signalled and the faulty section may bedisconnected. The risk of reverse currents that could result in damaging hot spots is reduced by the provision of diodes for each module and for each string of 10 modules.

The array support structure is illustrated in Figure 2.15. It has been designed to minimise foundation work at site and to facilitate erection.

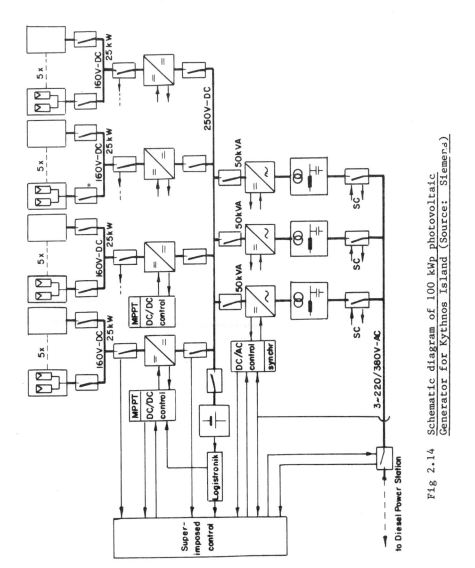

Fig 2.14   Schematic diagram of 100 kWp photovoltaic
Generator for Kythnos Island (Source: Siemens)

Lightning protection will consist of a system of overhead arrester wires. The metal frames of the modules, the array support structure and other metallic parts will be earthed.

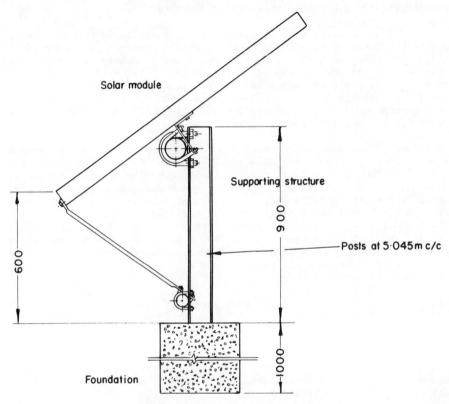

Solar module

Supporting structure

Posts at 5·045m c/c

600

9·00

1000

Foundation

Fig 2.15 Array support structure for Kythnos Island project (Source: Siemens)

2.4 Related technologies

Power conditioning

Apart from the simplest photovoltaic systems which directly use the dc power delivered by the array, some form of power conditioning is required. If batteries are incorporated to provide energy storage, a charge regulator is also needed to ensure that the battery is not overcharged or overdischarged. Some designers also favour the use of what is termed a Maximum Power Point Tracker (MPPT) or dc to dc converter, which ensures that at any given moment the maximum power is extracted from the photovoltaic array.

The basic requirement of a power conditioner is to convert the incoming dc power into the form required by the load, usually alternating current at the required voltage and frequency. The quality of the sinusoidal waveform must not only be adequate for the applied loads but also, if the system is operated in conjunction with the grid, meet the requirements laid down by the utility regarding harmonic content and power factor.

Inverters which perform this function have been available for many years for use with UPS (uninterruptible power supply) systems for computers, essential control equipment and for small wind energy systems. The requirements for the use of inverters with photovoltaic systems are more exacting in some respects than these existing uses and futher developments are in hand by several groups in USA and in Europe, stimulated by the various government and CEC-sponsored demonstration and pilot projects. The future market is potentially very large indeed.

The simplest inverters are externally commutated (ie, the power and frequency signal to generate the sine wave is taken from the grid connection). The models currently available do not in general provide ac power of high enough quality for their widespread use with grid-connected photovoltaic systems, particularly in respect of harmonic content and power factor. Stand-alone, self-commutated inverters can be designed to provide very high quality power but at present they are expensive and their reliability over a prolonged period in field conditions has yet to be demonstrated. Inverter operation when two or more are connected in parallel either in stand-alone mode or grid connected is an important area requiring definition and development.

Battery charge regulators have also been specially developed for use in photovoltaic systems, to ensure the optimum use of the relatively expensive power available from the photovoltaic array. Depending on the specific battery characteristics, the charge regulator will control the charging current in relation to the battery state-of-charge and protect the battery from being either over-charged, with consequent wasteful and potentialy dangerous gassing or over-discharged, which would reduce the life of the battery and impair its performance.

For some photovoltaic applications, the battery charge regulator is incorporated in a more complex load management control system, which directs the power available from the array to various loads in accordance with pre-established priorities. Power is fed to or drawn from the battery as necessary, subject to the control rules built into the load management device.

Some designers consider that the cost and additional complication of providing a maximum power point tracker (MPPT) is justified by the net gain in total power extracted from the photovoltaic array. A case can also be made for introducing several MPPT's, one for each section of the total array, to help minimise operational problems and losses due to module mis-match and partial shading of the array. At least for systems incorporating batteries (and thus the operating voltage is constrained within known limits), it seems clear that with good design, the net benefit to be obtained from the use of MPPT's is insufficient to justify their use, the extra energy obtained being only 2 or 3% over a year.

All power conditioning and load management systems are electronic devices which are either specially designed or are adapted from existing equipment.

They all involve some degree of parasitic power loss and their efficiency when operating at full and part loads is an important consideration. Some devices are very efficient at full load (over 95%) but have poor efficiency (sometimes less than 25%) at 10% load. Achieving a good part load efficiency is particularly important for systems which must operate for long periods at locations where the irradiance for much of the year is relatively low compared with clear day peak values in Summer. Such conditions frequently arise in Europe, as discussed in Chapter 3. Stimulated mainly by the requirements stipulated for the CEC-sponsored pilot plant programme, inverters and other power conditioning and control devices are now being developed in Europe that give over 90% efficiency down to 10% rated load.

One possible approach to achieving high efficiency over a wide range of loads is to use two or more power conditioning devices in parallel. When the load is small, only one unit operating near its rated load (and hence at high efficiency) is engaged. As the load increases, additional units are switched on, maintaining a high average efficiency. This approach however needs further development and demonstration, as a number of technical problems are involved.

It can be anticipated that steady progress will be made to identify reliable, cost effective designs of these electronic control devices. Although today rather expensive, often in excess of ECU 2.00/Wp for small systems, tremendous cost reductions may be anticipated with mass production of the designs which prove successful in the course of the next few years.

Energy storage

If electrical power is required when the sun is not shining or if a short-duration surge of power is needed to start electric motors or to meet an emergency situation, either some form of energy storage is needed or a back-up supply from the grid or another generator must be provided.

Batteries can be incorporated into photovoltaic systems but at present they are expensive, of the order of ECU 150 per kilowatt-hour stored. There are many different types of battery but for photovoltaic systems a lead-acid battery that is suitably designed for deep discharge and for a high-number of charge-discharge cycles is usually preferred. The best seem limited to about 1500 cycles, which limits the working life to considerably less than that of the rest of the system. Another important consideration is the battery efficiency under the particular operating conditions (ie, the ratio of useful power out to power in) and this in practice is often less than 85%. The use of batteries in a system will normally introduce an additional maintenance liability.

The critical factors for lead-acid batteries when used with photovoltaic systems are the number of charge/discharge cycles under non-ideal conditions and the sensitivity to deep discharge without being fully re-charged over a long period. Major research and development efforts are at present being directed in several countries towards the development of advanced batteries that are technically and economically suitable for powering electric vehicles. Much of this work will also be relevant for batteries to be used for photovoltaic systems but there are a number of differences which justify special efforts directed specifically to photovoltaic systems. For example, electric vehicle batteries must withstand much vibration and consequently the lead used in lead-acid batteries must be alloyed to provide adequate structural strength. Unfortunately all alloys used at present impair performance to some extent. For stationary photovoltaic installations, where vibrations in service do not arise, the possibility already exists for using pure lead or low antinomy batteries which have some significant performance advantages, such as lower self-discharge rate and higher in/out efficiency.

The US Department of Energy has for some years been sponsoring a growing battery research and development programme for specific solar applications. This programme has included experimental testing of battery systems at the Sandia Laboratories Photovoltaic Advanced Systems Test Facility and these tests are providing useful operational experience with a wide range of system types and sizes.

Several battery manufacturers have studied the special requirements of photovoltaic systems and can offer detailed advice on the choice of battery type and size. Varta (F R Germany) for example have prepared a detailed design guide and offer batteries and charge regulators specifically developed for photovoltaic applications (30).

Although improved lead-acid battery systems have in recent years become available for photovoltaic applications, the scope for further significant reductions in their capital cost and for improvements in their performance is limited. Several new types of battery have been identified in the USA as meriting further support, namely the iron-chromium Redox system being developed by the NASA Lewis Research Centre and the zinc-bromine flow-through system being developed by Exxon Research and Engineering Company. Gould Inc. are developing a zinc-bromine battery for utility 'peak lopping' applications and Battelle Columbus Laboratories are studying another system that shows promise, namely the nickel-hydrogen battery. In Australia, a battery said to be capable of storing 10 times more energy than lead-acid systems and needing no maintenance is being developed at the University of New South Wales. The battery has a lithium anode, a lead iodide cathode and a lithium/alumina electrolyte. The cost at present is prohibitively high.

There are several other options for energy storage besides batteries. One such is to employ a flywheel spinning at very high speed in a vacuum. The motor that drives the flywheel when surplus energy from the photovoltaic array is available can also be used as a generator to extract energy when required at night or during cloudy days. It has been reported that a system not much larger than a metre cube installed in a reinforced concrete housing could provide sufficient energy to supply a normal household for several days (31). New types of magnetic bearings and flywheel rotor materials are being investigated with the objective of developing reliable systems that could, with mass production, perhaps be competitive with advanced battery systems. Initial indications are promising but there is still far to go before such systems could be considered as a practical proposition.

Some applications for photovoltaic systems, such as water pumping or ice production, do not require electrical energy storage, since the product itself may be stored. Alternatively, photovoltaic pumping systems can be used in conjunction with an existing (or purpose built) pumped storage scheme or hydro-electric station.

One other energy storage system merits special mention, that is hydrogen. Photovoltaic systems are eminently suitable for producing hydrogen by electrolysis of water, and the hydrogen then liquified for use as a fuel in mechanical systems. This concept has been well described by Dahlberg (32), but is unlikely to be implemented until the use of hydrogen as a liquid fuel is generally accepted.

Lightning protection

The array field for a large photovoltaic system covers a large area and presents significant problems when considering lightning protection. A number of approaches have been followed for existing photovoltaic plants, depending on the degree of lightning risk and the level of protection deemed appropriate by the designer. The associated costs as a consequence vary widely.

2.5 Systems and applications

Introduction

The successful use of photovoltaic systems for powering space satellites led to increasing interest in developing photovoltaics for potential terrestrial uses. The growing concern regarding the finite supplies of fossil fuels and the environmental pollution associated with thermal power stations of all types has added impetus to the search for reliable and cost-effective photovoltaic power generators to suit a wide range of applications, from the smallest stand-alone systems to very large central power stations.

As the cost of the photovoltaic cells and modules falls and confidence in the reliablity of complete systems grows, an increasing range of applications will become economically viable in competition with conventional alternatives. An active commercial market is developing for various stand-alone systems incorporating photovoltaic generators. Completely self-contained photovoltaic generators incorporating battery storage are already competitive with small diesel generators for certain remote locations, even at today's prices. Indeed, it can be argued that photovoltaics provide the only realistic possibility for village pumping for water supply and general power supplies in remote parts of developing countries - the only factor limiting their general introduction now being the availability of the necessary finance. These systems offer social benefits out of all proportion to their cost, by improving the standard of living of the rural population and helping to arrest the migration to the towns of active young people.

Recognising the enormous potential for photovoltaics as a national energy resource and as an important industry in its own right, governments and industry have been active in developing and testing photovoltaic systems for various applications, small and large, with the aim of establishing reliable designs and reducing costs. A review of these activities is given below.

Stand-alone systems

There are currently many hundreds of relatively small stand-alone
photovoltaic systems operating throughout the world. Systems commercially
available include clocks, watches and calculators, battery chargers,
lighting units, highway warning signs, highway breakdown telephones, alarm
and security systems, pumps, refrigerators, automatic weather monitoring
stations, remote aircraft beacons and marine navigation aids (Figure 2.16).

Fig 2.16   Photovoltaic power for marine applications.
           This array provides power for internal lights,
           fog horn, gas and weather data transmitting
           equipment on a platform in the Gulf of Mexico
           (Source: Solarex Corp)

The first solar powered television set was installed as long ago as 1968 in
Niger.   Since then 123 schools in Niger have been similarly equipped and
several other countries have instituted similar programmes.   Photovoltaic
systems have proved reliable and cheaper to operate and maintain than the
large primary batteries or gasoline-engine generators previously used.

Small-scale solar powered pumping systems for irrigation applications have
long been considered to be of great potential importance for raising the
level of agricultural output in developing countries.   Studies have
indicated that systems pumping about 20 to 50m^3 /day from irrigation canals
and shallow wells would be appropriate for many millions of farmers who at
present either use human or animal power for irrigation or do not irrigate
at all.

In view of the great importance of raising agricultural production to feed
the world's growing population and to improve living standards, the United
Nations Development Programme (UNDP) in 1978 initiated a global project to
test and demonstrate small-scale solar powered pumping systems.   The UNDP

appointed the World Bank as executing agency and consultants were engaged to conduct a survey of existing technology, conduct field trials of systems that showed promise and carry out system design studies (33, 34). Phase I of the project has recently been completed with the conclusion that no solar pumping system at present available is directly suitable for widespread use for irrigation by farmers in developing countries, but that if improvements are introduced, the next generation of solar pumps could be technically appropriate. Total system costs are expected fall within 10 years to less than about ECU 5.00/Wp, when the unit cost of water pumped by such systems will be low enough to be generally economic for irrigation applications. However before they become economic for irrigation, solar pumps are likely to find wide acceptance for drinking water supplies in remote areas where the value of the water is greater.

Small pumping systems are particularly suited to photovoltaic power since the dc output from the array can be directly used to drive a dc motor coupled to a centrifugal pump. Electrical storage, with associated additional cost and complication, can be avoided since it is a simple matter to store the pumped water. More complex systems involving a multi-stage turbine pump or a reciprocating jack pump may be needed for deep well pumping. A low-head solar powered pumping system is illustrated in Figure 2.17 and a high head system in Figure 2.18.

Small solar pumps at today's prices are competitive with diesel pumps in remote locations where fuel supply problems are great and maintenance virtually impossible to provide. It was largely for these reasons that the international relief organisation Oxfam decided to purchase several solar pumps to serve refugee settlements in Somalia. It should also be noted that in recent years, the European Community, through the European Development Fund, has provided financial and technical assistance for the installation of more than 25 photovoltaic pumping systems for rural communities in Africa. Operating experience has been generally very good and further projects involving the installation of many more photovoltaic pumps are being planned.

Two other stand-alone applications are receiving increasing attention, namely refrigeration/ice making systems and desalination systems. The former is particularly important for preserving fish and other perishable foodstuffs and also for storing vaccines and other medical supplies. Photovoltaic powered refrigerators are already commercially available but further development is needed to extend the range, improve reliability and reduce overall system costs. As with pumping systems, refrigeration systems can be designed without the need for batteries since cold can be stored in the form of ice or in special eutectic compounds, having a high latent heat of freezing, and high insulation standards can be adopted.

It is important to stress that a photovoltaic system for a given application needs to be designed as a complete system and not as individual components. The photovoltaic generator should not be considered as a battery or diesel-generator substitute. All components interact and the objective of the design must be to obtain the most cost-effective combination of the overall system, taking into account efficiency, reliability and capital costs. Photovoltaic systems are particularly appropriate for applications where the final product itself may be stored, such as water or ice, rather than involve the additional cost and complexity of batteries. Usually it will be found that pumps and motors of

Fig 2.17   Solar powered low-lift pumping system on test
           in the Philippines
           (Source: Halcrow)

Fig 2.18   Solar powered high lift pumping system in Mali
           utilising a multi-stage borehole pump
           (Source: Leroy Somer)

higher efficiency than normally available will be more cost-effective, as will refrigerators with greater thickness of insulation.

In contrast to the relatively small stand-alone 'packaged' systems referred to above, there are many much larger stand-alone photovoltaic systems supplying power to such loads as individual houses, remote villages, large irrigation schemes and telecommunication installations. All of these installations constructed to date must be considered as experimental or pilot plants, built to enable equipment and systems to be developed and tested under real operating conditions. Brief details of some of the more interesting of these systems are given below as examples of the application of photovoltaic power.

San Hospital, Mali: The town of San, with 23000 inhabitants, is situated about 430km East of Bamako, Mali. The 100-bed hospital at San became the first in the world to be equipped with a solar generator when, in 1979, a stand-alone photovoltaic system was installed to provide all the electricity and water requirements (Figure 2.19). The electricity generator consists of a 8.9kWp, 120V dc array linked to 500Ah of battery storage and a 4kVA inverter to provide power at 220V, 50Hz ac for supplying medical apparatus, refrigerator, air-conditioning and lighting. There is also a 0.9kWp array supplying a vertical turbine pump in a borehole. Both arrays may be interconnected if required and an emergency diesel is also available in the event of breakdown or prolonged period of low insolation. Operating experience has been very good, the system having provided all the electricity and water needed by the hospital, with only minor maintenance being necessary. Financial and technical assistance was provided by the French Ministry of Cooperation and COMES (Commissariat à l'Energie Solaire). The main contractor was Pompes Guinard (France).

Picon, Western Java, Indonesia: A 5.5kWp photovoltaic power plant was supplied in 1980 for this village by AEG-Telefunken (F R Germany) as part of a research and technology agreement between the governments of Indonesia and F R Germany. The photovoltaic array is made up of 9 panels each with 72 modules (Figure 2.20). The tilt angle has been optimised to give maximum power in the dry season, when the photovoltaic generator will be mainly used for irrigation pumping to enable the villagers raise a second rice crop. The system includes a 60V, 420Ah battery and power conditioning equipment with inverter to provide 220V 50 Hz supply for all loads. The control system supervises the whole plant to ensure safe operation. Remote supervision is also possible through the transmission of the most important data via a radio link to Jakarta.

Mountain Refuge, Les Evettes, Savoie, France: As part of the CEC photovoltaic research and development programme, a 5kWp photovoltaic generator was designed and installed by Seri Renault in 1979 for a 66 bed hotel at an altitude of 2600m in the French Alps (Figure 2.21). The site is too isolated for the buildings to be connected to the grid. The 70m^2 array consists of 452 modules each of 11Wp arranged in 38 panels. A large amount of battery storage is provided to enable the normal daily load demand to be met for a period of up to 5 cloudy days. A 3kVA inverter is included in the power conditioning and control system, for supplying power to the refuge at 220V, 50 Hz. A comprehensive data recording system is also included, to enable performance to be evaluated.

Fig 2.19   San Hospital, Mali: 9.8kWp stand-alone photovoltaic system
(8.9kWp for electricity supply and 0.9kWp for pumping system
(Source: Pompes Guinard)

Fig 2.20   5.5kWp photovoltaic system for Picon, Western Java, Indonesia
(Source: AEG - Telefunken)

Fig 2.21   5kWp photovoltaic generator for alpine refuge,
           Les Evettes, France
           (Source: Seri-Renault)

Natural Bridges National Monument Park, Utah, USA: Rated at 100kWp, this large stand-alone system supplies all the electricity for two residences for park rangers, a dormitory for 20 summer-season workers, a visitors' centre and various maintenance areas (Figure 2.22). 600kWh of electrical storage is provided by batteries. Array output is 240Vdc and the inverter output is 240V, 60hz, single phase. The system also includes a diesel engine generator and battery charger for back-up service whenever the battery discharge reaches a specified low level. A high level of automatic control and load management has been incorporated in this system. Apart from a few relatively minor incidents, the system has functioned satisfactorily since installation in 1980.

Montpellier, France: A 26kWp, 120V dc stand-alone photovoltaic system was installed by France Photon on a farm near Montpellier in 1980 to provide power for an irrigation system sensing a land area of 600 ha (Figure 2.9). The irrigation system involves two stages of pumping. The first stage abstracts water from a river and delivers it to a reservoir. The second stage draws water from the reservoir for distribution at high pressure through the irrigation pipework. The performance of the system is being monitored by three organisations: INRA (Institut National de la Recherche Agronomique), IRAT (Institut de Recherche en Agromie Tropicale) and ENSAM (Ecole Nationale Supérieure Agronomique de Montpellier).

Grid-connected systems

Grid-connected photovoltaic systems installed for residences and industrial or commercial applications are designed to provide some or all of the power otherwise drawn from the grid. Usually surplus power generated by the photovoltaic array can be sold to the utility and fed back into the grid, with obvious economic advantages. At night or on cloudy days, power is drawn from the grid as required, thereby avoiding or reducing the need for battery storage. The design of a grid-connected system differs from a stand-alone system in two main respects:

- energy storage need not be incorporated; and

- the quality of the alternating current produced and fed into the grid must meet certain standards, particularly with regard to total harmonic distortion.

It is however generally true that it is not important for the design of the photovoltaic array whether the system is stand-alone or grid connected.

A number of grid-connected experimental plants have been built in the USA and elsewhere to develop systems and components and to obtain operating experience. Some of these are briefly described below.

Mount Laguna, California, USA: This system was commissioned in June 1979 and has a 60kWp, 230V dc array operating through an ac inverter to augment the existing diesel-generated electrical power system. There has been no overall system failure to date although the array has experienced considerable hot spot problems due to failed solar cells and modules. Use of bypass diodes and circuit redundancy has effectively prevented the hot spot problem from affecting overall system performance.

Mead, Nebraska, USA: This 25kWp facility began operation in Summer 1977 (Figure 2.23). It is a multi-purpose agricultural installation, providing power for irrigation pumps serving 35ha of corn 12 hours a day during the growing season. In early October of each year, the power is switched to drive fans in nearby corn-drying bins where a forced-air technique is used to dry the harvested corn. The system is also used to provide power for fertilizer generation. The array support structure can be adjusted to vary the tilt from 0 to 65° . 90kWh of battery storage is incorporated for pump starting and for power conditioning and data acquisition equipment. The system has proved very reliable in practice. A 10 to 15% degradation in array output has occurred over the three years of operation, but this is consistent with field experience elsewhere where early module designs with silicone rubber encapsulation systems have been used.

Bryan, Ohio, USA: A 15kWp, 128V dc photovoltaic system was installed in September 1979 at WBNO radio station in Bryan, Ohio, USA. The system provides power for a daytime AM radio transmitter requiring 4kW dc power. 40kWh of battery storage is also provided. The system was later modified to incorporate a dc-ac inverter and a load management system. The overall system has operated reliably from the start. On one occasion, the solar powered radio station was used by the police for communicating with members of their force when a utility failure shut down the main police transmitter.

Fig 2.22 Natural Bridges National Monument Park, Utah, USA
100kWp stand-alone photovoltaic system for park
residences, maintenance areas and visitors' centre

Fig 2.23 Mead, Nebraska, USA. 25kWp grid-connected
system for irrigation and crop drying
(Source: MIT-Lincoln Lab)

Residential systems

Studies in the USA have concluded that the most important grid connected
applications are those for private residences, mobile homes, low-rise
apartments and offices, small shops, schools, colleges, light industry,
administrative buildings and covered parking areas. None of these
applications is as yet economically viable but the US government is funding
a wide range of experiments in different parts of the USA to prepare for
the time when there will be a strong commercial market.

Regional experiment stations are being established in the northeast,
southwest and southeast areas of the United States. The need for three
separate sites arises from regional climatic differences. In the colder
climates, such as in the northeastern region, space heating could be
provided by the photovoltaic system either by conventional means or,
better, by an electric heat pump. Alternatively combined
photovoltaic/thermal collectors (PVT) or a combination of passive solar
heating and photovoltaics could be used. In the warmer climates, the
systems must be designed to produce additional electrical power for
air-conditioning equipment. The southeastern region has a hot climate with
high humidity levels and the southwestern region has a hot, arid desert
climate. The northeast and southwest residential experiement stations are
located in the Boston Massachusetts, and the El Paso, Texas, metropolitan
areas, respectively. A southeast station which had been scheduled to be
established in 1981 but has been delayed due to budget considerations.

Hardware development at each experiment station is in two stages. In the
first stage, prototype systems will be installed in the grounds of the
stations, with photovoltaic arrays on the roof of purpose-built buildings
containing the related power conditioning equipment. These prototypes are
unoccupied structures. The second stage will follow the successful
operation of a prototype system and it is expected that a refined version
of the same system will be installed in a house in the community near the
experiment station. The prototype systems will be carefully monitored for
performance and reliability, whereas the second stage systems will be used
to assess the reactions of the occupants.

A total of five prototype systems have been built and are currently being
evaluated at the northeast station. The first one, designed and built by
the MIT Lincoln Laboratory, is rated at 6.9kWp. The 93m^2 array is mounted
directly on the roof on a system of rollers and tracks to facilitate
installation and servicing. The other four systems include examples of
three types of roof mounting: stand-off, in which a structure supports the
modules above the roof weather sealing surface; direct, in which the
modules are placed directly on the roof; and integral, in which the
modules take the place of all roofing materials above the rafters.

Eight prototype systems are currently being designed for testing at the
southwest experiment station. Details of these and other residential
photovoltaic systems constructed or planned, are given in Table 2.1.
Particular mention should be made of the first lived-in photovoltaic
powered house, built in Carlisle, Massachusetts, USA (Figure 2.24). It has
a 7.5kWp photovoltaic array, mounted using a refined version of the MIT
Lincoln Laboratory system used at the northeast experiment station. The
house is passive solar heated, has solar thermal collectors for water
heating and incorporates many energy conservation features (35).

PROJECT	START CONSTRUCTION	OPERATIONAL	ARRAY	INVERTER	BUYBACK
Univ of Texas/Arlington	Jul 78	Nov 78	6.2 kWp G Sensor Tech	Gemini 8KVA	No
John F Long Fiesta	May 80	Jun 80	4.5 kWp D ARCO Solar	Gemini 8KVA	No
Carlisle, MA ISEE	Sep 80	Feb 81	7.3 kWp SO Solarex	Gemini 8KVA	TBD
Molokai Wiepke House	Mar 81	Apr 81	4 kWp SO ARCO Solar	Gemini 4KVA	TBD
Pearl City	Mar 81	Apr 81	4 kWp SO ARCO Solar	Gemini 4KVA	TBD
Kalihi	Mar 81	Apr 81	2 kWp SO ARCO Solar	Gemini 2KVA	TBD
Florida Solar Energy Center	Aug 80	Nov 80	5 kWp SO ARCO Solar	2-Gemini 4KVA	No
3 - NE ISEE	Sep 81	Jan 82			
4 - SW ISEE	Sep 81	Jan 82			
Tyndall AFB, FL. FPUP	RFP in preparation		2 kWp		
TVA FPUP (4 houses)	In project definition				
MIT/LL NE Prototype	Jul 80	Dec 80	6.9 kWp SO Solarex	Gemini 8KVA	No
GE NE	Sep 80	Mar 81	6.1 kWp D GE Shingle	Abacus 6KVA	No
Solarex NE Prototype	Sep 80	Feb 81	5.3 kWp SO Solarex	Abacus 6KVA	No
TriSolarCorp NE Prototype	Sep 80	Dec 80	4.8 kWp I ASEC	Gemini 8KVA	No
Westinghouse NE Prototype	Sep 80	Feb 81	5.4 kWp I ARCO Solar	Abacus 6KVA	No
TriSolarCorp SW Prototype	Jan 81	Apr 81	4.6 kWp I ASEC	Gemini 8KVA	No
Westinghouse SW Prototype	Feb 81	Apr 81	5.4 kWp I ARCO Solar	Abacus 6KVA	No
GE SW Prototype	Feb 81	Jun 81	5.6 kWp D GE Shingle	Abacus 6KVA	No
ASI SW Prototype	TBD	TBD	5.9 kWp D ARCO Solar	Gemini 8KVA	No
BDM SW Prototype	Jan 81	Apr 81	4.3 kWp SO Motorola	Abacus 6KVA	No
ARTU SW Prototype	TBD	TBD	4.9 kWp SO ARCO Solar	TBD	No
Solarex SW Prototype	Feb 81	Apr 81	4.8 kWp SO Solarex	Abacus 6KVA	No
TEA SW Prototype	Feb 81	Apr 81	3.7 kWp R Motorola	Abacus 6KVA	No

Array Code: G - ground mount   I - integral mount
             D - direct mount   R - rack mount
            SO - stand off mount

Table 2.1 Residential photovoltaic systems constructed or planned in USA

With residential systems also seen to be an important application in Europe, the Commission of the European Communities is sponsoring the construction of five 5kWp pilot plants to be built on a common site (to facilitate comparative evaluation) at Adrano, Sicily (Appendix B). These systems are expected to be in operation by the end of 1983.

Photovoltaic concentrator systems

A number of large photovoltaic concentrator systems have been built or are under construction in the USA. In addition, a 350kWp point focussing fresnel lens concentrator system has been built in Saudi Arabia to supply two villages as part of the joint USA/Saudi Arabia SOLERAS project. These concentrator systems have yet to be fully evaluated but already it seems clear that they will prove ultimately to be less cost-effective and less reliable than flat-plate photovoltaic systems. Details are given below of one major project to illustrate the complexity of the system and types of problem encountered.

A 320kWp photovoltaic concentrator system with combined thermal heating has been installed at the Mississippi County Community College, Blytheville, Arkansas, USA. It consists of 270 parabolic trough concentrators arranged in 45 rows of 6. The total aperture area is approximately 3780m^2. The collectors are mounted horizontally on the ground in a North-South orientation and track the sun East-West with ± 0.5% accuracy (Figure 2.25).

The receiver of the collector is V-shaped with a 25mm diameter pipe for coolant flow. The photovoltaic cells are mono-crystalline silicon attached to the outside surface of each side of the V. The top of the V is used for a cable tray for interconnection and instrumentation wiring, with a cover plate for protection.

The array electrical power is fed to a 400kVA inverter equipped with a maximum power point tracker. The inverter produces 480/277V, 60Hz, 3-phase ac power output, in parallel with the utility supplying the college. The utility has agreed to purchase all power generated in excess of college requirements at the same rate that power is supplied. The whole system has a central computerized data acquisition and control system.

A number of problems specific to concentrator systems have been experienced during commissioning of this system including poor thermal contact between solar cells and substrate, uneven illumination of cells, and birds roosting in collectors when in the stowed position with associated fouling of the reflector surfaces. As yet, the system has not been able to achieve its designed output.

Central utility generators

No large photovoltaic pilot plants of the size required for central generation by a main utility have yet been built, but plans are in hand for several in the USA, including one that will ultimately be 100MWp in size, to be built by the Sacramento Municipal Utility District in California. Agreement to proceed with the construction of a 1MWp plant for Southern California Edison has recently been announced by ARCO Solar Corporation

Fig 2.24   The Carlisle house, Massachusetts, USA.
           The first lived in photovoltaic house,
           with a 7.5kWp photovoltaic array
           (Source: MIT-Lincoln Lab)

Fig 2.25 Mississippi County Community College, Arkansas, USA
View of collectors for 320kWp concentrator photovoltaic
generating system combined with thermal heating
(Source: Mississippi County Community College)

(USA). In Europe, the largest scheme announced so far is the 1MWp DELPHOS project in Italy. There appear to be no major technical problems associated with multi-megawatt central photovoltaic generators, since such systems, being modular, can be based on smaller systems of proven design, although further research and development is needed to establish optimum configurations.

Another concept for central power generation is the Space Power Station (SPS), in which a very large photovoltaic array is constructed in space where it is constantly illuminated by high intensity direct solar radiation. The array would be in geostationary orbit and the output power would be continuously beamed to an earth receiving station by means of a micro-wave link. Studies undertaken in the USA indicate that the concept is technically feasible and would generate power at competitive prices, but the practical and political problems would appear to be formidable (36). Such a scheme would in any case not be acceptable near populated regions and may be discounted for Europe.

European pilot plant programme

Anticipating that photovoltaic power systems will in time become an economically viable energy resource for Europe as well as an important export for European industry, the CEC has recently initiated a comprehensive programme of pilot plant installations, covering a wide variety of applications. Power ratings range from 30kWp to 300kWp, as listed in Appendix B. A summary of the main features of these pilot plants is given in Table 2.2. Design work is already well advanced on all systems and construction is expected to start in mid-1982, for completion by June 1983.

2.6 Prospects for cost reduction

Modules and systems

Reference has been made several times in this chapter to the prospects for future cost reductions for various processes and systems. Whilst it is clear that progressive cost reductions have occurred in the past, it is important to distinguish between what may with some confidence be expected in future and what is largely speculative. From the time that major development work into photovoltaics for terrestrial applications began in the late 1960's and early 1970's, manufacturers and government organisations have been studying the prospects for reducing costs, through improved techniques and higher volume production, and producing cost forecasts and targets. One such study, carried out by CNES in France in 1974, compared the module price forecasts of manufacturers in USA and France as part of a much wider study (37). The resulting curves are shown in Figure 2.26 (note 1974 currency values). Experience since 1974 has shown that these forecasts were remarkable accurate.

In 1980, the US Department of Energy published revised cost goals for photovoltaic modules and systems for the period 1975-1990, as shown in Figure 2.27. Although current module prices are somewhat higher than the target, current system costs are within the target range.

Table 2.2 Principal features of EC photovoltaic pilot plants (sheet 1 of 2 sheets)

Pilot Plant No	Contractors	Location	Application	PV Array Power kWp	Battery Size Ah	Back-up System	Surplus Power to Grid	DC-DC Converter % Efficiency 10% load	full load	DC-AC Inverter % Efficiency 10% load	full load	Size and Type of Inverter (see key)	PC & C Cost ECU/kWp
1	AEG-Telefunken	Pellworm F R Germany	Recreation centre	300	8000	Grid	Yes	Not used		83	92 / 95	3 X 75kVA acf / 1 X 300kVA bcg	800
2	Siemens, PPC, Varta	Kythnos Island Greece	Supply to island network	100	2400	No	Yes	96	97	82	94	3 X 50kVA ade	750 Note 1
3	ENEL, Ansaldo, Solaris Varta	Alicudi Island Italy	Village	76	500-600 kWh	Diesel gen	No	Not used		>80	>90	1 X 40kVA ade	1000
4	Lucas BP	Marchwood UK	Supply to grid	80	370	No	Yes	Not used		70	93	1 X 100kVA acf	1850
5	ACE, Ansaldo, Petrochemical Indelet	Tremiti Islands Italy	Desalination plant	64	2250	No	No	>94	>95	80	94	Several adgj	750 Note 1
6	IDE-ACEC, Fabricable, Photowatt	Chevetogne Belgium	Swimming pool	63	275kWh	Grid	Yes	Not used		92	95	6 X 10kVA adej	810-1250 Note 2
7	Seri Renault, Jeumont Schneider, Varta	La Guyane Indian Ocean	Village	60	1500	Diesel Gen	No	Not used		90	93	2 X 40kVA adej	611 Notes 2,3
8	Photowatt, Oldham Batteries	Mont Bouquet France	TV and FM transmitters	50	800	Grid	No	Not used		85	90	1 X 30kVA ade	1296 Notes 2,5
9	Photowatt, Oldham Batteries	Nice France	Airport systems	50	800	Grid	No	Not used		85	90	1 X 30kVA ade	1296 Notes 2,5

Table 2.2 **Principal features of EC photovoltaic pilot plants (sheet 2 of 2 sheets)**

Pilot Plant No	Contractors	Location	Application	PV Array Power kWp	Battery Size Ah	Back-up System	Surplus Power to Grid	DC-DC Converter % Efficiency 10% load	full load	DC-AC Inverter % Efficiency 10% load	full load	Size and Type of Inverter (see key)	PC & C Cost ECU/kWP
10	University College Cork, AEG, ESB	Fota Island Cork Ireland	Dairy farm	50	600	Grid	Yes	Not used		88	96 / 95	3 X 10kVA adfgj / 1 X 50kVA bcfj	805
11	Holec, AEG	Terschelling Island Holland	School	48	500	Grid	Yes	83	96	NA	94	1 X 60kVA bce	1100
12	Seri Renault, PPC, Jeumont Schneider, Varta	Aghia Roumeli Crete Greece	Village	50	1500	Diesel gen	No	Not used		90	93	1 X 40kVA adej	767 Notes 2,3
13	Pragma, AGIP Nucleare	Giglio Island Italy	Ozone and cold store	43	64 kWh & cold	Grid	No	90	97	NA / NA	95 / 95	1 X 4kVA ade / 1 X 12kVA acej	650
14	Leroy Somer, France Photon, Oldham Batteries	Paomia-Rondulinu Corsica France	Village	42	3000	Diesel gen	No	Not used		>77	>87	1 X 50kVA ade	1330 Note 4
15	Gensun-C50	Hoboken Antwerp Belgium	Hydrogen plant and pumping	30	Hydrog & water	Grid	No	Not used		Not used		Not used	–

Key: a self commutated d transistors g square wave output
b line commutated e PWM h single phase
c thyristors f wave form synthesis j eff excl transformer

Notes:
1 Serial production 4 700 for serial prod
2 Inverter only 5 1160 for serial prod
3 500 for serial prod 6 Project now reduced to 30 kWp

- 57 -

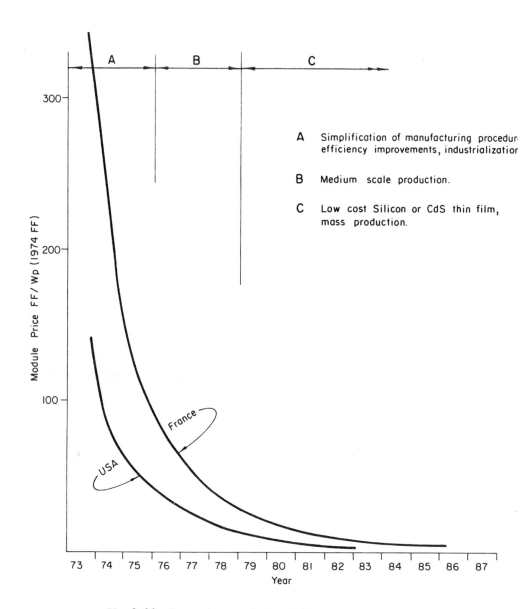

Fig 2.26 Comparison made in 1974 of forecast
photovoltaic module costs in USA
and France (1974 FF)

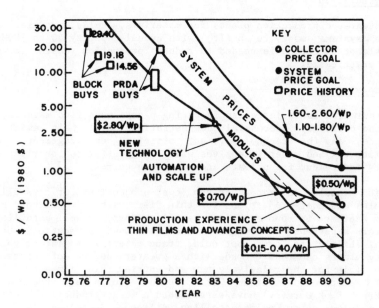

Fig 2.27   US DOE draft cost goals for photovoltaic
modules and systems (1980 $)

Until recently, the emerging photovoltaic industry in the USA had been
strongly supported by the US government. Cost targets had' been
consistently met each year, but the budget cut backs by the Reagan
administration will have a major impact on the US photovoltaic research,
development and demonstration programme. It is now generally accepted that
the cost targets may in consequence be delayed by several years. Costs
will no doubt continue to decrease in real terms, but not as quickly as
previously hoped.

European prices for photovoltaic modules and systems have until recently
been somewhat higher than for similar products in the USA, due mainly to
smaller production volumes. This situation is rapidly changing as new
manufacturing plants in Europe come on stream. In mid-1982, FOB module
prices for large volume orders were being quoted down to ECU 7.00/Wp from
certain manufacturers in Europe and in USA. This price is equivalent to
about ECU 6.00/Wp when adjusted to 1980 currency values.

Although the cost targets for modules and complete sytems are based on
detailed analyses of large scale production techniques, and may thus be
considered as technically feasible given the assumptions, what is actually
achieved will be strongly dependent on three main factors:

1. Technical progress

    Progress towards achieving cheaper materials and process must continue
    as anticipated, and this is largely dependent on the level of public
    and private funding of research and development.

2. Industry build-up

Large integrated production plants for materials, cells, modules and complete systems must be built, with annual capacity 50 to 100MWp. This scale of operation is needed to reduce production costs to the minimum practicable.

3. Market development

The output from these large manufacturing plants must find markets, in Europe and overseas. Successful demonstrations will help establish confidence in the technology, but official encouragement and appropriate tax and other incentives will be essential to provide the necessary inducement for private and public investment.

A distinction should be made between cost projections for crystalline silicon modules and cost projections for thin film systems. Silicon cell technology, whether in mono or poly-crystalline form, is well established with many years experience of production and use. Cost projections based on the availability of lower cost solar grade material and larger volume manufacturing plants can thus be made with a greater degree of certainty than is possible for the relatively new and untried thin film approaches, whether for amorphous silicon or cadmium sulphide techniques. AEG-Telefunken (F R Germany) have established a large production facility for making modules using polycrystalline silicon (Silso) cells and, assuming their Phase II and Phase III extension schemes go ahead as planned, expect to be able to reduce prices to DM 1 to 3/Wp (ECU 0.40 to 1.24/Wp) by 1985, as indicated on Figure 2.28.

A useful survey of photovoltaic system cost experience in the USA has been prepared by Burgess et al (38). The costs of six photovoltaic flat plate systems were broken down into eight cost account categories. A typical ground mounted system was then compared with the costs of three lower cost systems obtained by:

1. taking a system using all the best design features from among the six systems;

2. taking a mid-term system based on costs represented by what is expected to be possible when US $2.80/Wp modules become available; and

3. taking a long-term system based on costs represented by what is expected to be possible when US $0.70/Wp modules become available.

The results of this interesting exercise are shown in Table 2.3. The main conclusions can be summarised as follows:

- Current flat plate systems range in price from $20 - $30/Wp (ECU 18.9-28.3/Wp).

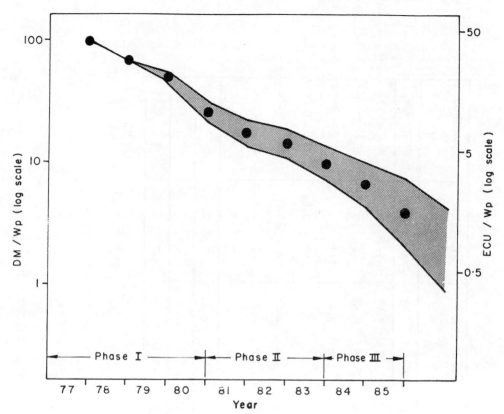

Fig 2.28   <u>AEG – Telefunken cost projection for photovoltaic modules</u>

- This price is dominated by engineering design and module costs.

- Using the best design features from current demonstration projects would give a BOS cost of $4.50/Wp (ECU 4.25/Wp).

- Using standard modular designs and low cost design features would give a BOS cost of $2.19/Wp (ECU 2.07/Wp).

- Mid-term systems could be obtained for $6.06/Wp (ECU 5.72/Wp).

- Long-term systems could be obtained for $2.35/Wp (ECU 2.21/Wp).

The conclusions are encouraging in that they lend support to the DOE cost targets (but not, of course, saying when these might be achieved).

With the above discussion in mind, it has been possible to prepare Figure 2.29 which is an updated cost projection for photovoltaic modules and systems, based on the latest information from manufacturers and trends in the industry. A considerable degree of uncertainty has to be included in the cost projection due to the dependence on the three factors referred to earlier. The main features of the cost projection, referred to as the CEC 1982 cost projection, are as follows:

SYSTEM SIZE (kWp)	CURRENT 100 6.7		"BEST" CURRENT 300 7.0		MID-TERM 300 8.5		LONG-RANGE 300 10	
ARRAY EFFICIENCY (%)	¢/m²	¢/Wp	¢/m²	¢/Wp	¢/m²	¢/Wp	¢/m²	¢/Wp
Engineering	560	8.41	88	1.25	57	0.75	29	0.32
Structure	44	0.66	46	0.66	42	0.55	42	0.46
Foundation	70	1.05	22	0.32	(1)	(1)	(1)	(1)
Civil Work	137	2.05	67	0.96	18	0.24	18	0.20
Electrical	173	2.59	50	0.71	20	0.26	20	0.22
Pwr. Cond. & Ctrl.	180	2.70	32	0.46	26	0.31	13	0.13
Buildings	38	0.57	7	0.10	7	0.08	7	0.07
BOS Sub-Total	1202	18.03	312	4.46	170	2.19	129	1.39
PV Modules (2)	753	11.32	716	10.23	297	3.87	88	0.96
TOTAL SYSTEM	1955	29.35	1028	14.69	467	6.06	217	2.35

(1) Subterranean foundation, cost in structure and civil work

(2) FOB price plus distribution and mark-up

Table 2.3 Photovoltaic system cost comparison

- 62 -

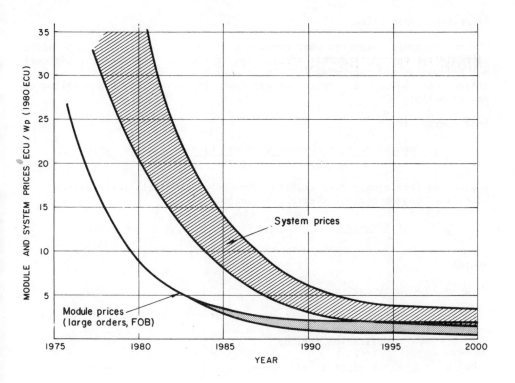

Fig 2.29 CEC 1982 cost projection for modules and systems (1980 ECU)

Year	Modules FOB ECU/Wp	Systems Installed ECU/Wp
1980	9	20 – 40
1985	3.00 – 3.50	8 – 14
1990	1.00 – 2.00	3 – 6
1995	0.70 – 2.00	1.8 – 3.8
2000	0.50 – 2.00	1.6 – 3.5

This cost projection can now be used to calculate the electricity unit cost from a photovoltaic generator.

Electricity unit price

The key economic parameter when comparing alternative means of generating electricity for any specific application is the unit energy cost, the cost per kilowatt-hour. Assuming the full annual electrical output of the system is used, the unit energy cost is given by the following relationship:

unit energy cost =

$$\frac{\text{annual capital charges + annual operating costs}}{\text{total energy output}}$$

Neglecting inflation, tax credits and other financial incentives, a simplified approach gives the relationship:

$$p \quad = \quad \frac{r.C + m.Wp}{\eta_{ave}.Ea}$$

where

p = unit cost per kWh

r = interest plus amortization factor (depends on discount rate and amortization period)

C = total capital cost of system

m = specific operating cost, expressed as cost per peak Watt installed

Wp = total installed peak power in Watts

η_{ave} = annual average conversion efficiency of system

Ea = total annual solar energy incident on the array.

As an example, let us consider a 10kWp grid connected system built in southern Europe in the late 1980's when photovoltaic module costs are expected to have fallen to about ECU 2.00/Wp and total installed costs for a complete generator system including some battery storage to about ECU 5.00/Wp. The total capital cost would thus be ECU 50000. Assuming a 20 year amortization period and 5% discount rate, the factor r would be 0.0802 and thus the annual capital charges would be ECU 4012. Annual operating cost including maintenance and insurance, might be of the order of ECU 0.06/Wp and thus the total annual operating cost would be ECU 600. Taking the total area of a 10kWp array to be 100 square metres and assuming the installation was at a place where the total annual solar energy incident on the plane of the array was 1750kWh/m^2, Ea would thus be 175000kWh. Based on an annual average conversion efficiency of 7.5%, the total annual energy output would be 13125kWh. The unit energy cost would thus be:

$$p = \frac{4012 + 600}{13125} = \text{ECU } 0.35/\text{kWh}$$

It should be noted that the interest rate used in the above example is the real rate of return given by the difference between the cost of money and the inflation rate. If the required real rate of return is taken as 10% instead of 5%, the unit energy cost in the above example would be ECU 0.49/kWh instead of ECU 0.35/kWh.

This simplified approach to photovoltaic system economic analysis is represented graphically on Figure 2.30, which can be used to derive unit energy cost for different values of capital cost, interest rate, specific operating cost, total incident solar energy and average system conversion efficiency.

The next step is to consider the electricity unit cost given by alternative sources for comparison with the photovoltaic system, to see when the break-even points occur (39, 40). For example, a typical price in Europe for grid electricity supplied to domestic consumers is ECU 0.08/kWh. A typical price for electricity generated by large (ie, 1-10MW) diesel or gas turbine generators (GTs) in remote areas is about ECU 0.20/kWh. For small

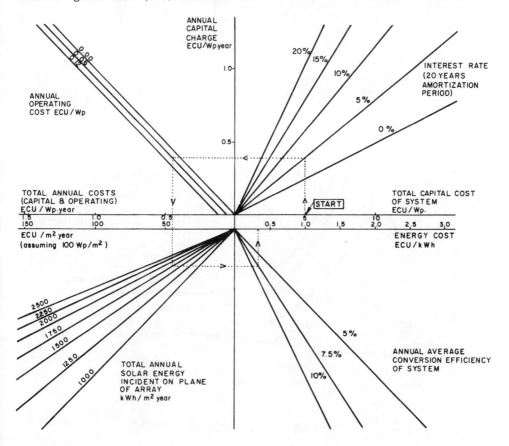

Fig 2.30   Simplified approach for calculating unit energy cost

(ie, 10-1000kW) diesel generators in remote places, the cost can often exceed ECU 0.50/kWh. These unit prices may be expected to rise at a rate higher than the general inflation rate. In the 1970's, the differential inflation rate for commercial energy in most countries was over 10% per annum, but it is generally considered unlikely that such a high rate will obtain in future. The probable range for the differential inflation rate for electricity supplied from conventional sources over say the next 20 to 50 years is between 5 and 10%.

Thus, whereas the price of photovoltaic systems and hence the cost of the electricity generated by such systems is expected to fall in real terms over the next 10-20 years, the cost of electricity from conventional generators is almost certain to rise in real terms over that period. The consequences are illustrated in general terms in Figure 2.31, which shows that photovoltaic systems in regions with high solar insolation (eg, southern Europe) could be competitive with small diesels in remote areas by the mid 1980's, with larger diesels and GTs in remote areas by the late 1980's and with grid supplies by the mid 1990's. For places with less solar insolation (eg, central and northern Europe), the corresponding break-even dates would be later by some 5 to 10 years or longer.

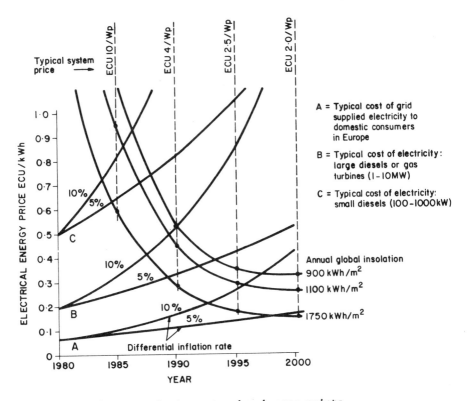

Fig 2.31 Photovoltaic system break-even points

- 66 -

Clearly, for specific systems and locations considerably more detailed economic analyses need to be made, taking into account other factors such as taxation and inflation, but the above simplified approach provides a good indication of the potential for photovoltaics to be competitive with conventional generating systems, including in time grid supplied electricity.

CHAPTER THREE - PHOTOVOLTAICS AS AN ENERGY RESOURCE FOR EUROPE

3.1 European climate

Introduction

The potential for photovoltaic systems in any situation is largely dependent on two controlling factors:

- the cost of reliable photovoltaic devices and systems

- the climate, in particular the solar irradiance.

We have seen, in Chapter 2, how photovoltaic technology has developed in the last 20 years from very specialised and expensive systems for space satellites to a point where there is already an active commercial market for certain small-scale applications such as cathodic protection, telecommunications and packaged lighting units. Attention is now being directed to the development of larger systems, providing power for remote houses, villages and rural industry. In a few years, system costs are expected to fall to a level where photovoltaics will be used economically in conjunction with grid supplies and eventually for large-scale central generation plants.

Clearly photovoltaic generator systems will find their first application in places where (a) the cost of electricity provided by alternative means is high, and (b) the climate is favourable. The main opportunity will arise first for remote areas in countries in and near the tropics, but there is expected to be a growing place for photovoltaics as an energy resource for non-tropical regions, including Europe.

Geographic factors have a strong influence on the amount of solar energy that reaches the earth's surface. When designing a solar powered system it is necessary to consider climatic factors such as global and diffuse solar irradiance, the daily and seasonal variation of irradiance, the associated ambient temperatures, wind conditions and cloud patterns. Local factors can strongly influence the climate. It is also important to consider periods when the actual conditions depart widely from long term averages.

Insolation and latitude

The terms used for solar radiation quantities are irradiance (power per unit area), irradiation (energy per unit area) and insolation (energy per unit area in a specified time period), which are explained in more detail in the Glossary. Figure 3.1 shows how much solar energy is received on average on the horizontal plane in various parts of the world. Figures 3.2 and 3.3 indicate the seasonal variation.

The mean global insolation is greatest in the continental desert areas between latitudes 30° S and 30° N (the world's 'solar belt'), exceeding 6.8kWh/m^2 per day (2500kWh/m^2 per year) in some places. It is somewhat less at many places near the equator due to cloud and considerably less in high latitudes due to the solar altitudes being low and to cloud. It is

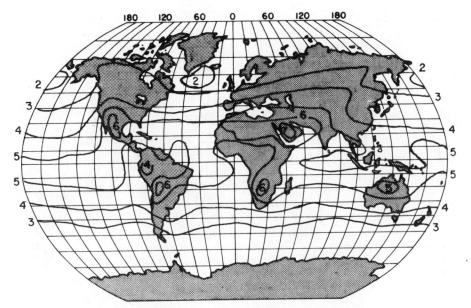

Fig 3.1 Annual mean global insolation, kWh/m² per day

somewhat less at many places near the equator due to cloud and considerably less in high latitudes due to the solar altitudes being low and to cloud. It is significant to note that northwest Europe receives about 1300kWh/m² per year, compared with about 1750kWh/m² per year in southern USA and over 2200kWh/m² in sahelian Africa and the Middle East. The world insolation maps (Figures 3.1, 3.2, 3.3) do not attempt to show regional variations, which can be quite marked particularly in mountainous areas. Figure 3.4 shows contours of annual mean insolation for the member countries of the European Community. Figures 3.5 and 3.6 indicate the seasonal variation for the same area, which is more marked in the northerly latitudes. These diagrams are largely based on data published in the European Solar Radiation Atlas, vol I, published by the CEC (41). Data for Greece are based on information presented in references 42 and 43. Although detailed climatic records have been kept by practically all countries for many years, there is a distinct lack of long term solar radiation data, there simply being no demand for such data until recently. The European Solar Radiation Atlas is an important document that brings together such data as are available for Europe and, after careful processing, presents the results in a consistent and convenient form. Work is currently in hand to extend the scope of the Atlas to include Greece and other European countries not at present represented.

In view of the great potential for photovoltaic power foreseen in the United States of America, it is of interest to compare average insolation values there with those for Europe and the world as a whole. Figure 3.7 shows contours of mean annual insolation for continental USA and Figures 3.8 and 3.9 indicate the seasonal variation. Meteorologically, continental USA may be divided into three macro-climatic regions. The southwest is characterized as very dry and hot with short winters, long summers and

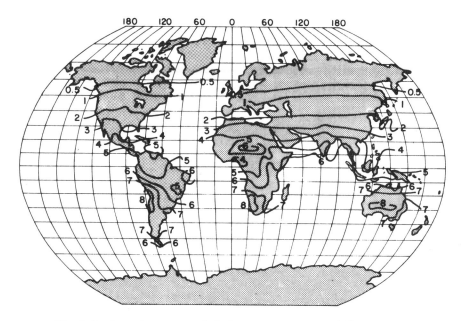

Fig 3.2 <u>December mean global insolation, kWh/m^2 per day</u>

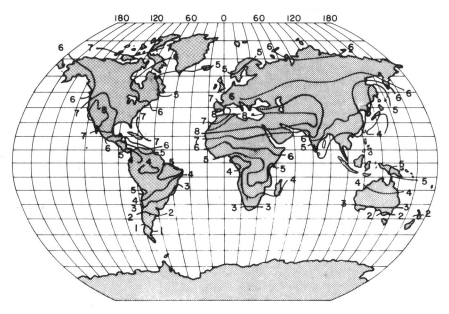

Fig 3.3 <u>June mean global insolation, kWh/m^2 per day</u>

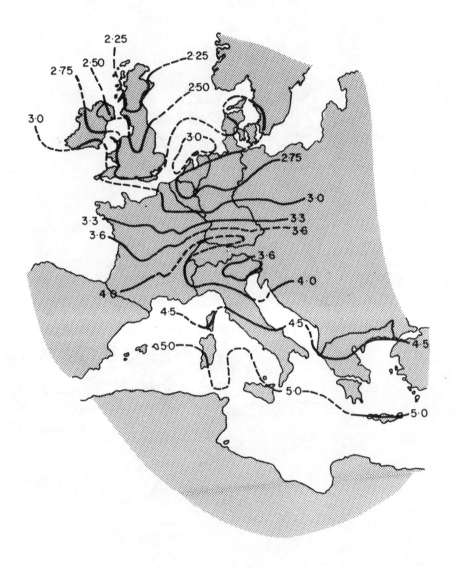

Fig 3.4 Annual mean global insolation for Europe, kWh/m^2 per day

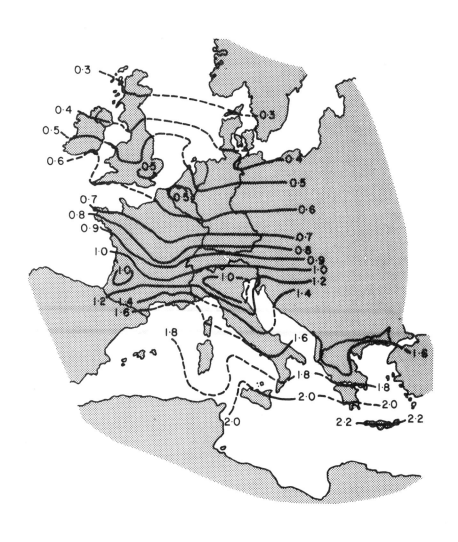

Fig 3.5 <u>December mean global insolation for Europe, kWh/m^2 per day</u>

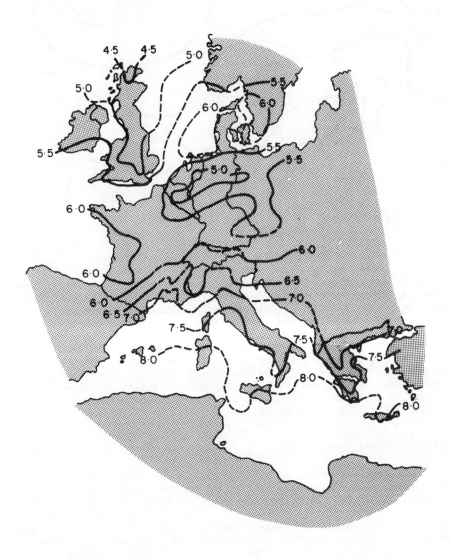

Fig 3.6 June mean global insolation for Europe, kWh/m^2 per day

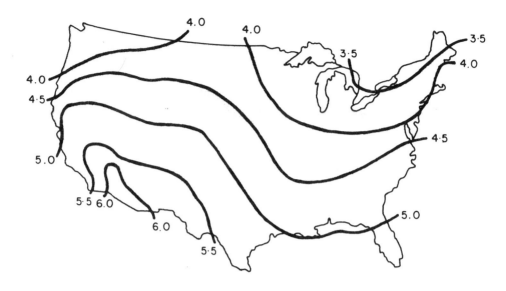

Fig 3.7 <u>Annual mean global insolation for USA, kWh/m² per day</u>

usually clear skies. The southeast is tropical with high humidity, moderate winters and summers and partly cloudy skies. The northeast has a temperate climate with long cold winters and short mild summers. Skies are often overcast in winter and vary between clear and partly cloudy in summer. Most of western Europe is classified as having a temperate wet climate. Summer on the Mediterranean shores is distinctly dry and hotter than the more northerly and westerly areas which also have a less well-defined rainy season. The mountainous central regions are colder but again have a distinct dry season. The central Iberian plateau is classified as desert. Cloudy skies are very frequent along the Atlantic coastal belt and over the mountains.

Global and diffuse insolation

The seasonal variation of insolation is an important factor since it affects the amount of energy storage or back-up generation needed for a particular photovoltaic installation. Another important factor is the magnitude of the diffuse component of the global insolation and its variation throughout the day. Diffuse radiation cannot in general be optically focussed, and therefore photovoltaic concentrator systems are not suitable for places having a high proportion of diffuse radiation, whereas flat plate photovoltaic arrays can convert both direct and global insolation.

The degree of seasonal variation of global insolation and the diffuse component varies widely. Treble (3), using data provided by the UK Meteorological Office, illustrates this clearly by comparing the monthly mean insolation at Kew, London (Lat 51.5° N) with that for Aden, People's Democratic Republic of Yemen (Lat 12.5° N) as shown in Figures 3.10 and 3.11. The diffuse element is shown hatched and the direct element clear.

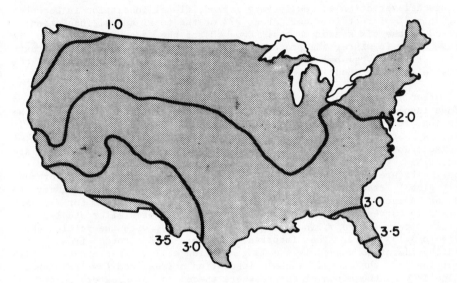

Fig 3.8 December mean global insolation for USA, kWh/m^2 per day

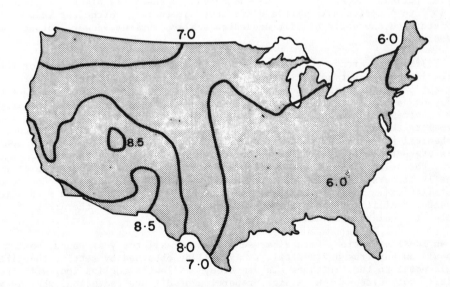

Fig 3.9 June mean global insolation for USA, kWh/m^2 per day

The seasonal variation at Kew is very marked, global insolation in December being only 7% of that in June. About 77% of the total annual insolation is received in the six summer months. Some 60% of the annual insolation is diffuse and in winter the proportion can be as high as 70%. By contrast, the lowest monthly global insolation at Aden is 75% of the maximum and the diffuse element is about a third of the global.

Since flat plate photovoltaic arrays are able to convert both direct and diffuse insolation, for design purposes it is necessary to know only the variation in global irradiance on the plane of the array. A particularly useful way of presenting irradiance data is in the form of irradiance/duration curves, as derived by Bourges (44). Some examples are selected here to indicate the differences between three locations in Europe. Figure 3.12 shows the curves of cumulative frequency of mean hourly global irradiance on a horizontal plane for the months January to June at Ajaccio, Corsica. Figure 3.13 shows similar curves for Nice and Figure 3.14 curves for Hamburg. ENEL have carried out similar studies for several sites in Italy, including Messina, Sicily, with the results shown in Figure 3.15 (45). The important feature to note from these irradiance/duration curves is the long duration of low irradiances. A high proportion of the total annual insolation arises from low irradiances, particularly in the north of Europe where there is a greater number of cloudy days than in the south. Thus, it is important to ensure that photovoltaic systems are capable of operating at high efficiency not only at times of high irradiance but also at low levels.

Tilt angle

So far in this discussion, we have been concerned only with irradiance on a horizontal surface. If the photovoltaic array is turned towards the sun, the direct irradiance is increased in accordance with the cosine of the angle between the solar beam and the normal to the array plane. Angles up to 15° in practice make little difference but it is nevertheless important to determine the angle of tilt needed to achieve maximum power output of the array.

By tilting the array, the amount of diffuse irradiance on the plane of the array is reduced but normally this is offset by the effect of ground reflection. This can be a substantial bonus if the ground surrounding the array has a highly reflective surface such as concrete or snow. Clearly the incident irradiance may be maximised by tracking the sun, keeping the normal to the array plane directed towards the sun. This introduces mechanical complexity with associated maintenance demands. Not only does this present practical problems, particularly in developing countries, but also the reliability of the system is adversely affected. Therefore it is usually preferable to fix the array plane at the optimum angle. On some installations, the tilt angle may be seasonally adjusted by hand — this is a useful facility provided the operators remember to re-set the array tilt angle at regular intervals.

In an ideal situation, with clear sunny weather all the year round, maximum annual output from a fixed-tilt array will be obtained by setting the tilt angle equal to the latitude. In practice, a lower inclination will be better for sites with a high proportion of diffuse radiation, whereas a steeper angle will increase output during sunny winter days and thus reduce

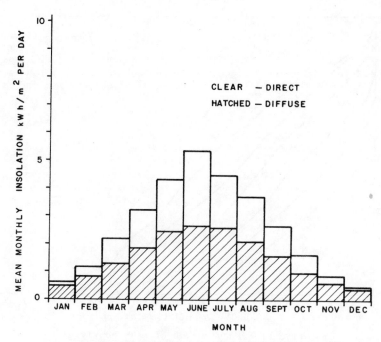

Fig 3.10   Variation of monthly mean insolation at Kew, UK

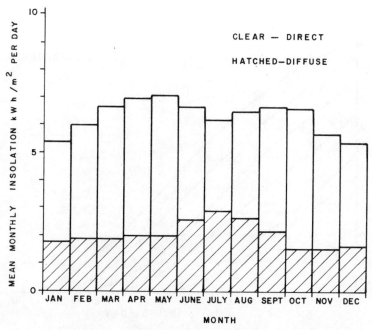

Fig 3.11   Variation of monthly mean insolation at Aden,
           People's Democratic Republic of Yemen

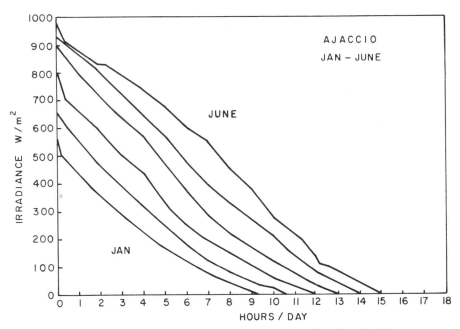

Fig 3.12   Cumulative frequency curves of mean hourly global
           irradiance on a horizontal plane for Ajaccio, Corsica

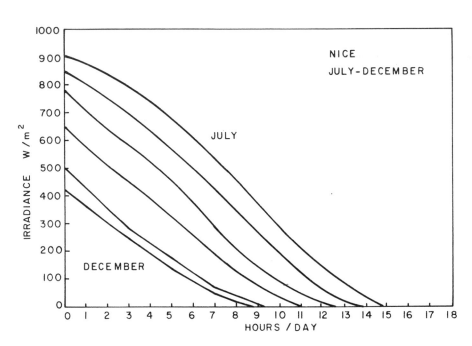

Fig 3.13   Cumulative frequency curves of mean hourly global
           irradiance on a horizontal plane for Nice

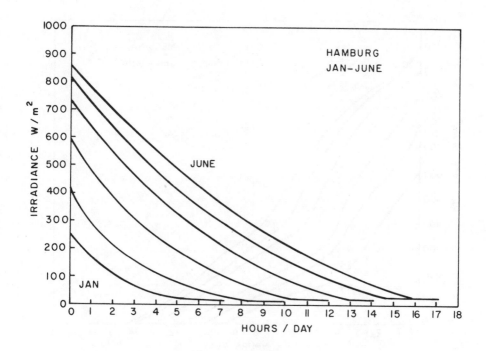

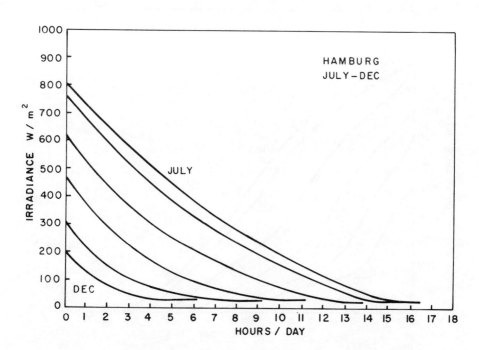

Fig 3.14 Cumulative frequency curves of mean hourly global
irradiance on a horizontal plane for Hamburg

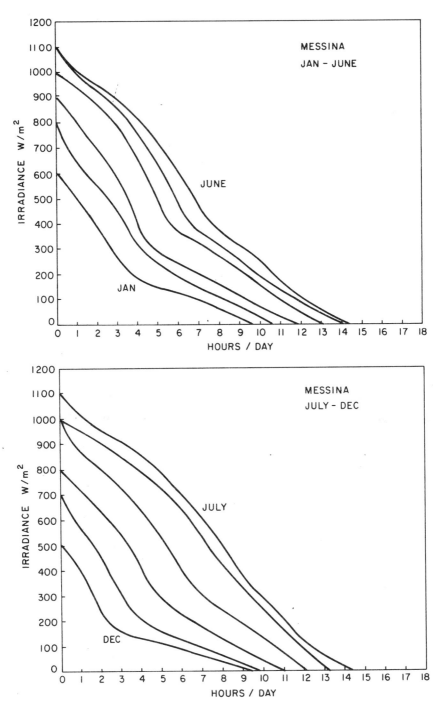

Fig 3.15 Cumulative frequency curves of mean hourly global irradiance on a horizontal plane for Messina

storage or back-up generator requirements.

The analysis required to optimise tilt angle is quite complex, since insolation data is available in general only for a horizontal plane (see for example reference 46). In many cases also, only global monthly mean values are available and it is then necessary to estimate the diffuse proportion. Statistical methods may then be applied to derive the daily and hourly variation (47).

A number of computer programs are now available for determining the optimum tilt angle for a particular site. As an example of such a prediction the results are presented here of the calculations made by Adriatica Componenti Elettronici in collaboration with the Universita dell'Aquila Istituto di Fisica Tecnica, Italy, in connection with the CEC-sponsored 65kWp pilot photovoltaic plant to be built on the Tremiti Islands (48). The mean global insolation on a horizontal plane at this site is 1217kWh/m^2 per year and Figure 3.16 shows the calculated variation of global insolation on an inclined plane facing south. The optimum fixed tilt angle is about 32 degrees (less than the latitude which is 42° N) giving a global insolation on that plane of about 1350kWh/m^2 per year, an 11% increase over the value for a horizontal plane.

Figure 3.17 shows the effect of varying the tilt angle each month. The results are summarised in Table 3.1, which shows that varying the tilt each month would produce a global insolation of 1404kWh/m^2 per year in the plane of the array. Since this is less than 4% greater than the insolation obtained for a fixed tilt of 32 degrees, the additional complication is not justified.

Size of photovoltaic system

The size of the photovoltaic system needed to provide a given amount of energy per year is dependent on the location. The most important parameter for a flat-plate fixed array is the global insolation on a horizontal plane. The ambient temperature will also have some influence, as the efficiency of solar cells decreases with rising temperature. Ground reflection effects may give a significant boost to systems in areas subject to snowfall - provided the snow is not allowed to build up on the array surface itself.

As a rough guide to how location affects array sizing, Table 3.2 gives the approximate array area needed at various locations for a system to produce 1000kWh electrical output per year. The calculation assumes that the tilt of the array is fixed at its annual optimum value and that the global insolation in the plane of the array exceeds the global insolation on a horizontal surface by the following amounts:

Latitude (N or S)	Increase in insolation in plane of array
0 - 20	0
20 - 35	5%
35 - 50	10%
50 - 65	15%

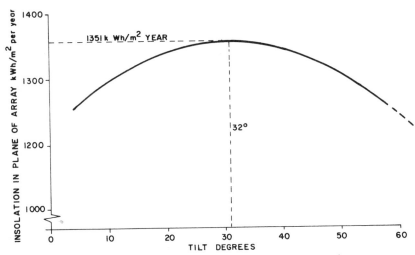

Fig 3.16 Insolation in plane of array for fixed
array tilt angle, for Tremiti islands

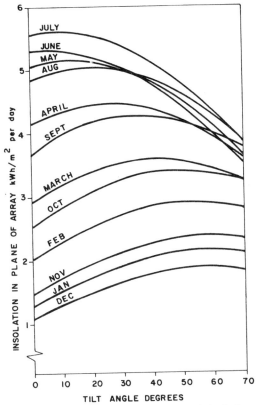

Fig 3.17 Effect of varying tilt angle on global insolation
in plane of array for each month, for Tremiti islands

For the purpose of deriving peak power rating, the specific output of the array was taken to be $100Wp/m^2$, a figure comparable with typical photovoltaic modules available today.

Month	Optimum tilt angle degrees	Average insolation in month kWh/m^2
Jan	60 – 65	67.4
Feb	50 – 55	81.4
March	40 – 45	110.6
April	25 – 30	134.0
May	10 – 15	159.4
June	5 – 10	159.1
July	5 – 10	173.1
August	20 – 25	155.6
Sept	35 – 40	128.2
Oct	50 – 55	105.4
Nov	60 – 65	70.9
Dec	60 – 65	58.9
	Total	1404.0 kWh/m^2 per year

Table 3.1   Insolation in plane of array with optimum tilt angle each month, for Tremiti Islands

Location	Latitude degrees	Mean global insolation on horizontal plane in kWh/m² per year	Mean global insolation in plane of array in kWh/m² per year	Array area in m² for 1000 kWh electrical output per year for overall system conversion efficiency			Array peak power in watts for 1000 kWh output per year for conversion efficiency		
				5%	7.5%	10%	5%	7.5%	10%
NW Europe									
London	51	912	1049	19.1	12.7	9.5	1910	1270	950
Brussels	51	942	1083	18.5	12.3	9.2	1850	1230	920
Amsterdam	52	1086	1249	16.0	10.7	8.0	1600	1070	800
Copenhagen	55	1028	1182	16.9	11.3	8.5	1690	1130	850
Dublin	53	1072	1233	16.2	10.8	8.1	1620	1080	810
Central Europe									
Paris	48	1127	1240	16.1	10.8	8.1	1610	1080	810
Bonn	50	1006	1107	18.1	12.0	9.0	1810	1200	900
Luxembourg	49	1065	1172	17.1	11.4	8.5	1710	1140	850
S Europe									
Rome	41	1684	1852	10.8	7.2	5.4	1080	720	540
Trapani	38	1915	2107	9.5	6.3	4.7	950	630	470
Athens	38	1752	1927	10.4	6.9	5.2	1040	690	520
Madrid	40	1643	1807	11.1	7.4	5.5	1110	740	550
N America									
N USA	45	1500	1650	12.1	8.1	6.1	1210	810	610
SE USA	30	1800	1890	10.6	7.1	5.3	1060	710	530
SW USA	32	2200	2310	8.7	5.8	4.3	870	580	430
S Canada	50	1300	1430	14.0	9.3	7.0	1400	930	700
Central America									
West Indies	20	1800	1800	11.1	7.4	5.6	1110	740	560
Mexico	10	1800	1800	11.1	7.4	5.6	1110	740	560

Table 3.2 Approximate array size for photovoltaic system producing 1000 kWh per year at different locations (Sheet 1 of 2)

Location	Latitude	Mean global insolation on horizontal plane in kWh/m² per year	Mean global insolation in plane of array in kWh/m² per year	Array area in m² for 1000 kWh electrical output per year for overall system conversion efficiency			Array peak power in watts for 1000 kWh output per year for overall system conversion efficiency		
	degrees			5%	7.5%	10%	5%	7.5%	10%
S America									
Brazil	8	1800	1800	11.1	7.4	5.6	1110	740	560
Argentina	30	1500	1575	12.7	8.5	6.3	1270	850	630
Chile	30	1500	1575	12.7	8.5	6.3	1270	850	630
Asia									
Japan	35	1500	1575	12.7	8.5	6.3	1270	850	630
China	35	1500	1575	12.7	8.5	6.3	1270	850	630
East Indies	5	1700	1700	11.8	7.8	5.9	1180	780	590
Philippines	15	1800	1800	11.1	7.4	5.6	1180	740	560
India	20	1500	1500	13.3	8.9	6.7	1330	890	670
Middle East	30	2400	2520	7.9	5.3	4.0	790	530	400
Central USSR	60	1100	1265	15.8	10.5	7.9	1580	1050	790
Africa									
Sahelian	25	2300	2415	8.3	5.5	4.1	830	550	410
Equitorial	0	1800	1800	11.1	7.4	5.6	1110	740	560
Southern	30	1800	1890	10.6	7.1	5.3	1060	710	530
Australasia									
Australia	30	1800	1890	10.6	7.1	5.3	1060	710	530
New Zealand	45	1100	1210	16.5	11.0	8.3	1650	1100	830

Table 3.2 Approximate array size for photovoltaic system producing
1000 kWh per year at different locations (Sheet 2 of 2)

Well designed systems using present technology should give an annual average overall conversion efficiency of 7.5%. For this performance, the array area required would be about 13m^2 in London, 11m^2 in Amsterdam, 7m^2 in Rome and Athens, and only 6m^2 in Trapani, in southern Italy. As array area and associated peak power rating is the main factor determining total system cost, it is apparent that photovoltaic systems giving similar annual energy output would cost nearly twice as much in northern Europe as they would in southern Europe. This difference, although considerable, does not necessarily mean that photovoltaic systems will not find economic applications in northern Europe, since it is necessary to consider the cost of the alternatives and other less tangible factors when assessing the potential for photovoltaics, as discussed later in this chapter.

3.2 European energy scene

The problem

The oil crisis in 1973/74 initiated a fundamental change in the ability of industrialized nations to control the ordered development of their economies and thereby secure their future. Previously only major wars, revolutions and deep economic depressions had so directly affected the lives of ordinary citizens in the world's oil importing nations. The insecurity of the world's energy supplies has subsequently been demonstrated with the substantial disruptions to oil shipments and continuing uncertainty resulting from the recent political changes in Iran and the war between Iran and Iraq.

The pressure has been felt even more acutely in developing countries which must divert ever increasing amounts of their limited foreign exchange resources to pay for oil imports. They cannot easily pass on the extra cost to those consumers unable to afford them, nor has it proved possible for the additional foreign exchange to be obtained by increased prices for their primary exports. The foreign exchange restraint has compounded the problems by making it more difficult for them to maintain existing energy consuming plant and to invest in new plant. The result has been the virtual paralysis of the economies of some nations. The ordinary citizens are very aware of the problem, as they experience frequent power cuts and long queues at fuel pumps (Figure 3.18).

All countries, either individually or in groups formed out of common interest, are seeking solutions to their energy problems and a plan for the future. The European Community has been collaborating for some time with the International Energy Agency (IEA) and the International Institute for Applied Systems Analysis (IIASA) in the field of energy supply and demand and its relationship with the economies of member countries. The first report attempting to quantify the European Community's long-term energy future has recently been published (49) and it finds that Europe is in an exceptionally difficult position in terms of solving its energy problems. Europe is heavily industrialized with an essential need to export in order to maintain and improve living standards. This results in a high specific energy demand, but its accessible fossil fuel resources cannot meet demand. Coal reserves appear to be large but much of these require deep mining which is costly. Uranium resources are also limited and the continued

Fig 3.18   The oil crisis affects the motorist

expansion of nuclear generating capacity has associated technical and social problems. New and renewable energy resources, such as direct solar, geothermal and wind energy, will certainly not be able to supply a major proportion of Europe's total estimated energy needs by the year 2000, due to the long lead times and capital investment associated with new technologies, even when they are technically proven and commercially viable. Substantial decoupling between Gross Domestic Product growth and energy demand growth is thus essential if economic development is not to be hampered by the modest rate of growth in energy supply which is achievable over the next 40 to 50 years.

A further important factor is that the European Community is a group of sovereign nations and not, as yet, a political unity. In comparison with say the USA or the USSR the lack of an overall government in Europe slows the development of new energy technologies in a co-ordinated manner and by the same token reduces the political influence Europe would otherwise expect to have in the world in relation to its population and economic wealth.

The energy policy recommendations of the CEC try to take account of this situation by identifying a number of practical strategies which may be summarised as follows:

- a steady reduction in the European Community's dependence on energy imports

- a diversification of energy imports by type and source

- the improvement of relations between energy exporting countries and the European Community

- world-wide collaboration in seeking comprehensive solutions to the energy problems.

The increasing requirement for the introduction of new and renewable sources of energy is illustrated in Figure 3.19, which is based on information presented in reference 49. Total energy demand by 2025 is seen as growing to between twice or three times the present level. As oil supplies diminish, so new energy supplies must increase to keep pace with the growing demand in Europe. The 3% growth rate in energy demand may be considered unrealistically optimistic in terms of developing new sources quickly enough: even meeting a 2% growth in demand will be a considerable task, requiring huge investments in renewable and conventional energy resources. Coal and nuclear energy resources could perhaps be expanded at a rate of 4% per year and the requirement then for renewables would be for an expansion rate of 5.5% per annum total, similar to the historic growth rate of the energy supply between 1960 and 1973. There is little scope for further development of hydro resources in Europe and the proportion of the total energy demand in the year 2000 to be met by renewables (excluding hydro power) would still only be a few percent. However the studies reported in reference 49 lead to the conclusion that Europe should aim to achieve 10% of energy supply from renewable resources by 2025 and a few percent of supply achieved by 2000 is an essential precursor to this target.

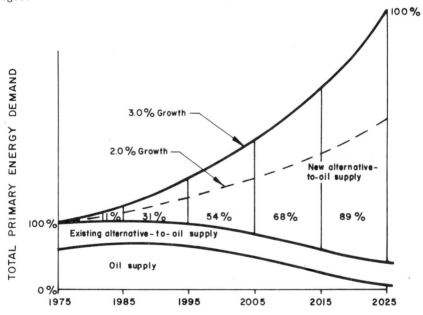

Fig 3.19   Timing and magnitude of the energy problem

The implications of these figures for solar photovoltaic energy are of great importance. Hydro, biomass and marine energy resources have considerable physical limitations. Geothermal resources are limited by the available geological formations and engineering capabilities. The limits on the direct use of solar, biomass and wind resources are set primarily by climatic, energy storage and land availability constraints. A broad review of renewable energy resources and their potential is being studied by many organisations (see for example references 50 and 51), but the scale of the problem as it relates to photovoltaics may be illustrated quite simply: assume the total energy demand in Europe in the year 2025 is 1380 mtoe, (ie, assume 2% growth rate). Assume that renewables account for 10%, namely 138 mtoe, and that the solar photovoltaic energy provides 15% of energy from renewables, say 20 mtoe. This amount of oil would generate about 80 x 10^9 kWh/year, requiring on the basis of current technology a total installed peak capacity of photovoltaic generators of over 60GWp. To achieve this order of magnitude would require an annual average growth rate of photovoltaic installations of over 30% per year starting from a base of 1MWp in 1985 - a high but not totally impossible goal, provided the right decisions are made soon.

Energy and electricity statistics

Before proceeding to consider specific market areas for photovoltaics, it will be useful first to review the statistics relating to electricity production and energy in Europe. This can best be done by a series of tables and diagrams based on statistical data published in Eurostat 1980 (52) and the Energy Statistics Yearbook 1980 (53).

Table 3.3 lists the area, population, population density and estimated population growth for the European countries and for the USSR, USA, Canada and Japan.

The growth in Gross Domestic Product and the main aggregated energy functions from 1975 to 1979 for the European Community (9 countries) is shown in Figure 3.20. The growth in total energy consumption for the same period is shown in Figure 3.21. The gross inland energy consumption per capita and the gross industrial energy consumption per capita are shown in Table 3.4.

Three tables summarise the production and consumption of electrical energy for the year 1978: Table 3.5 lists the maximum output capacity of power stations; Table 3.6 lists the net production of electrical energy; and Table 3.7 lists the consumption of electrical energy. It is of interest to note that western Europe is a net importer of electricity from power stations located in eastern Europe.

Finally, Figure 3.22 shows the trend in the total net production of electrical energy over the 10 years 1970 to 1979 for each of the European Community countries. The rate of growth is at present virtually static in several countries as a result of the combined effect of conservation measures and economic recession. This has resulted in over-capacity in some countries.

These figures and tables provide the context in which the scope for solar photovoltaic generators in Europe is considered.

Country	Area '000 sq km	Population '000	Density per sq km	Projected population '000	
				1985	1990
EUR 9	1 525.6	259 774	170	260 858	263 203
1 F R Germany	248.6	61 327	247	59 614	58 587
2 France	544.0	53 277	98	54 829	56 085
3 Italy	301.3	56 714	188	57 078	57 830
4 Netherlands	41.2	13 942	338	14 250	14 648
5 Belgium	30.5	9 840	323	9 840	9 887
6 Luxembourg	2.6	358	138	358	360
7 United Kingdom	244.1	55 902	229	56 164	56 844
8 Ireland	70.3	3 311	47	3 538	3 718
9 Denmark	43.1	5 104	118	5 187	5 244
10 Greece	132.0	9 361	71	9 477	9 696
11 Spain	504.8	36 781	73	39 050	40 642
12 Portugal	91.6	9 796	107	10 206	10 471
13 Turkey	814.6	43 210	53	51 225	57 502
14 Norway	323.9	4 059	13	4 133	4 175
15 Sweden	450.0	8 278	18	8 356	8 350
16 Switzerland	41.3	6 337	153	6 284	6 315
17 Austria	83.9	7 508	90	7 454	7 436
18 Finland	337.0	4 758	14	4 858	4 923
19 USSR	22 402.2	261 260	12	279 558	291 637
20 USA	9 363.1	218 059	23	232 880	243 513
21 Canada	9 922.3	23 499	2	25 490	26 826
22 Japan	377.6	114 898	304	119 732	122 769
World	135 830.0	4 258 000	31	4 827 476	5 272 667

Table 3.3 Area, population, density per square kilometre and estimated population growth (mid 1978)

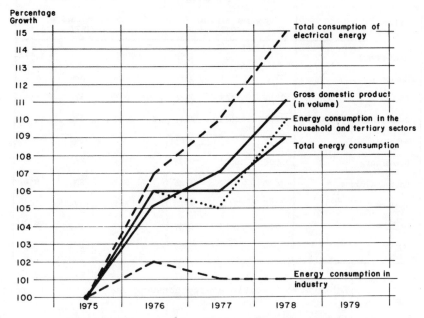

Fig 3.20 Energy consumption growth by sectors

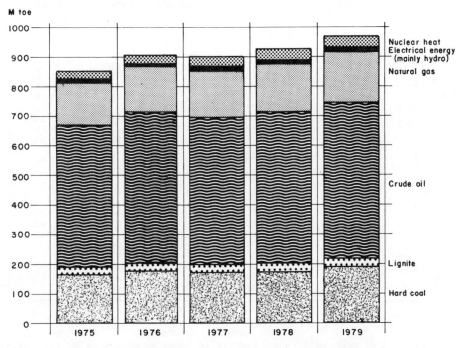

Fig 3.21 Total primary energy consumption (EUR 9)

	Gross inland energy consumption kgoe per capita			Industrial energy consumption kgoe per capita		
	1976	1977	1978	1976	1977	1978
F R Germany	4231	4200	4359	1047	1011	1039
France	3202	3141	3317	816	828	837
Italy	2310	2247	2274	683	679	639
Netherlands	4777	4633	4647	862	860	928
Belgium	4753	4526	4783	1348	1322	1385
Luxembourg	10903	10606	10694	7499	6969	6868
United Kingdom	3672	3772	3746	841	838	816
Ireland	2026	2144	2222	461	524	503
Denmark	3719	3851	3746	550	595	637
EUR 9	3498	3475	3559	870	861	863

Table 3.4 Specific energy consumption

	Country	Hydro MW	Geothermal MW	Nuclear MW	Thermal MW
	E U R 9	45 066	398	23 716	219 995
1	FR Germany	6 403	–	7 776	66 215
2	France	18 577	–	6 439	29 244
3	Italy	15 388	398	1 113	26 561
4	Netherlands	–	–	497	16 410
5	Belgium	499	–	1 670	8 060
6	Luxembourg	1 213	–	–	221
7	United Kingdom	2 446	–	6 221	64 542
8	Ireland	532	–	–	2 283
9	Denmark	8	–	–	6 459
10	Greece	1 476	–	–	3 390
11	Spain	12 735	–	1 073	12 920
12	Portugal	2 480	–	–	1 270
13	Turkey	1 880	–	–	2 990
14	Norway	17 400	–	–	160
15	Sweden	13 800	–	3 742	7 810
16	Switzerland	10 980	–	1 021	600
17	Austria	4 496	–	–	7 817
18	Finland	2 390	–	1 080	6 495
19	USSR	47 550	6	9 616	178 300
20	USA	70 300	560	50 900	432 000
21	Canada	41 950	–	5 495	26 855
22	Japan (a)	27 317	150	13 399	91 045

(a) Situation at 31.3.1979

Table 3.5 Maximum output capacity of power stations in MW (end 1978)

Country	Total generation	Net Production (a)				
		Hydro	Geo-thermal	Nuc-lear	Conven-tional thermal	Total
EUR 9	1 187 433	140 916	2 384	115 159	865 159	1 123 618
1 F R Germany	353 450	18 204	–	33 856	280 500	332 560
2 France	226 692	68 537	–	28 999	119 716	217 252
3 Italy	175 041	47 138	2 384	4 159	113 733	167 414
4 Netherlands	61 596		–	3 811	55 151	58 962
5 Belgium	50 838	496	–	11 872	35 988	48 356
6 Luxembourg	1 389	311	–	–	1 007	1 318
7 United Kingdom	287 689	5 194	–	32 462	231 148	268 804
8 Ireland	9 978	1 013	–	–	8 416	9 429
9 Denmark	20 780	23	–	–	19 500	19 523
10 Greece	21 050	2 978	–	–	16 766	19 744
11 Spain	99 534	41 007	–	7 302	46 927	95 236
12 Portugal	13 942	10 532	–	–	3 048	13 580
13 Turkey	21 726	9 185	–	–	11 277	20 462
14 Norway	80 997	80 101	–	–	129	80 230
15 Sweden	92 901	57 074	–	22 718	10 532	90 324
16 Switzerland (b)	43 904	32 510	–	7 995	1 845	42 350
17 Austria	38 069	24 608	–	–	12 321	36 929
18 Finland	35 800	9 744	–	3 106	21 133	33 983
19 USSR	1 201 900	168 000		41 500	926 000	1 135 500
20 USA	2 442 800	284 001	2 900	276 403	1 722 111	2 285 415
21 Canada	346 160	234 190	–	29 435	72 712	336 337
22 Japan (c)	567 500	74 187	1 000	56 131	406 022	537 340
World	7 537 000	1 510 000	6 500	583 00	5 030 000	7 130 000

(a) i.e. after deduction of the amount taken by station auxiliaries.
(b) Year 1.10.77 – 30.9.78
(c) Fiscal year 1.4.78 – 31.3 79

Table 3.6 **Production of electrical energy (1978)**

Country	Total net pro- duction	Net imports	Con- sump- tion for pumping water	Transp. and distr. losses	Consumption of the internal market
EUR 9	1 123 618	+ 13 204	8 224	73 832	1 054 766
1 F R Germany	332 560	+ 3 085	2 027	12 370	321 248
2 France	217 252	+ 4 288	753	15 934	204 853
3 Italy	167 414	+ 2 126	2 765	15 204	151 571
4 Netherlands	58 962	+ 345	–	2 749	56 558
5 Belgium	48 356	– 2 777	368	2 571	42 640
6 Luxembourg	1 318	+ 2 533	356	110	3 385
7 United Kingdom	268 804	– 76	1 429	21 812	245 487
8 Ireland	9 429	–	526	1 009	7 894
9 Denmark	19 523	+ 3 680	–	2 073	21 130
10 Greece	19 744	+ 127	–	1 367	18 504
11 Spain	95 236	– 1 532	1 783	9 562	82 359
12 Portugal	13 580	– 218	70	1 400	11 892
13 Turkey	20 462	+ 621	–	2 115	18 968
14 Norway	80 230	– 3 405	225	7 316	69 284
15 Sweden	90 324	– 1 008	27	7 857	81 432
16 Switzerland (a)	42 350	– 5 394	1 361	3 131	32 464
17 Austria	36 929	– 2 762	559	2 523	31 085
18 Finland	33 983	+ 1 278	–	2 261	33 000
19 USSR	1 135 500	– 17 500	–	98 000	1 020 000
20 USA	2 285 415	+ 19 442	5 900	204 485	2 094 472
21 Canada	336 337	– 19 510	1 900	29 640	285 287
22 Japan (b)	537 340	–	4 594	28 493	504 253
World	7 130 000	–	650 000		6 480 000

(a) Year 1.10.77 – 30.9.78
(b) Fiscal year 1.4.78 – 31.3.79

Table 3.7 Consumption of electrical energy (1978)

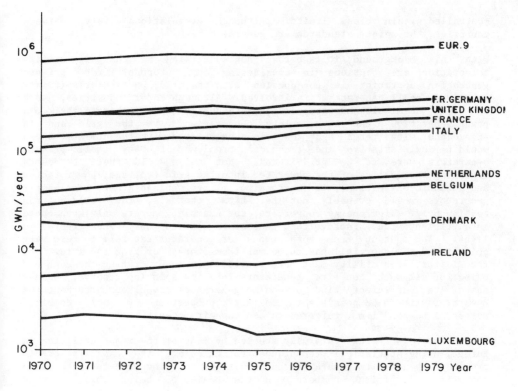

Fig 3.22 Trend in the total net production of
electrical energy (log scale)

Electricity generation

Electricity generation in Europe is characterized at present by large
monopoly utilities, mostly state owned, which have responsibility for the
generation and distribution of electrical power throughout the country (or
major portion of the country). The central power plants under construction
or built recently are generally large, of at least 1000MW installed
capacity, to obtain maximum economy of scale. The lead time for the
planning, design, construction and commissioning of such plants is at least
7 years for oil or coal stations and often over 10 years for nuclear
plants.

The transmission and distribution of electrical power through national
grids is practically universal in mainland areas. Only the more remote,
sparsely populated mountain areas and small islands are without grid
supplies. In most of these parts, small privately-owned diesel generators
are used to supply villages or individual homes.

Consumers in Europe have come to expect a high degree of reliability from
their grid supplies. The utilities assure this standard by maintaining a
high proportion of spinning reserve to cope with unexpected breakdowns or
load increases. The quality of the power supplied is also strictly
maintained, with voltage, frequency, power factor and harmonic distortion

controlled within close limits. Although regulations vary between countries, the safety standards are generally high.

With this background, it is perhaps not surprising that the large utility corporations are cautious in considering any proposal for private generation operating in conjunction with the grid, particularly if many relatively small generators are involved. Although major industries, such as steel, chemicals and paper-making, frequently own and operate their own power generation plant in complete conjunction with the grid and in accordance with terms and conditions negotiated with the utility, there would be administrative and technical problems if many small private generators were to be grid-linked. Not only would there be doubts regarding the quality of the power fed into the grid (voltage, power factor and harmonic distortion), it must also be recognised that most small generators would probably not be 'firm' that is, they would not run continuously. Wind energy converters, for example, operate only when there is sufficient wind. Photovoltaic generators operate only when there is light. The output from both types of generator can fall to zero very quickly and the shortfall has to be met from storage or from the grid. If the latter, the utility must make provision by incorporating the combined output of all such 'non firm' generators into its spinning reserve, unless there are sufficient wind or solar generators dispersed throughout the country for it to be possible to consider a proportion of their combined output as being 'firm', referred to as 'capacity credit'.

This issue is being carefully studied by many of the major utilities in Europe, using statistical techniques to calculate the capacity credit provided by non-firm generators such as solar and wind when connected into the grid in sufficient numbers, on the basis of being able to meet anticipated peak demand on the system with a specified reliability (ie, a given low probability of failure). In general, solar generators offer better capacity credit than wind generators, since the solar input is better matched to the typical diurnal load curve on a grid system. For example, studies carried out by the Italian Utility ENEL indicate that the introduction of each 1MW of solar generation into a large network would allow a reduction of nearly 0.3MW of conventional generating capacity, if the total solar penetration does not exceed a few percent. In other words, the capacity credit is about 30%. In the ENEL studies (see, for example, references 54 and 55), it should be noted that the capacity credit is more precisely defined in financial terms as:

Capacity Credit ($/kWh) =

$$\frac{\text{Decrease in Capital Charge (\$ per year)}}{\text{Energy Output of Solar Plant (kWh/year)}}$$

The decrease in capital charges arises from the lower conventional capacity needed in a generating system when a solar plant is introduced into that system.

Economic analysis of the effect of adding photovoltaic systems to the grid is a complex yet important task, since it is only by such analysis that the full economic implications can be assessed. The technical implications for the control of generation, transmission and distribution throughout a major grid network incorporating many relatively small photovoltaic generators are also complex. In some respects, for modelling purposes it is sometimes possible to treat photovoltaic generators as negative loads, to some extent predictable but essentially uncontrolled

Although there will undoubtedly be difficulties, it is nevertheless clear that the utilities in Europe as elsewhere are beginning to take increasing account of the need for electricity to be generated using renewable energy resources including wind and solar. This will often involve privately owned generators working in conjunction with the grid, giving rise to a number of technical and administrative problems; indeed in some countries, private generation of electricity for sale, either to the utility or to other consumers, is at present prohibited by law. Legislation may be needed to stimulate both the utilities and and consumers to make progress in agreeing terms and conditions regarding the use of private generators connected with the grid.

In the United States, such a stimulus was provided by the Public Utility Regulatory Policies Act (PURPA) which was passed into law in November 1978. PURPA regulations require all electric utilities to sell electricity to and purchase electricity from private power producers, such as owners of photovoltaic generator systems. The utility must also take into account any requirements for back-up power. The rates for buying and selling such power have in general yet to be determined by the various state utility commissions, but PURPA specifies that the buyback rates shall be reasonable and reflect the 'avoided costs' to the utility of not having to generate and transmit the electricity. High buyback rates, say at least 50% of the selling price of electricity, will clearly provide a strong boost for privately-owned photovoltaic systems. The California State Energy Commission now requires utilities to buy back excess power from customers with photovoltaic generators at 100% of the selling rate for electrical power, although this is being disputed.

In Europe, a number of countries have active programmes to introduce wind energy converters of steadily increasing size. These programmes are generally being carried out by or with the full involvement of the electricity utility and so far the question of private generation and buyback rates has not often arisen. In Denmark however it is understood that the utility has recently agreed to a buyback rate of 30% for a wind energy converter to be installed soon by a private operator. In the United Kingdom, the principle of reasonable buyback rates based on avoided costs is already established for private generators of all types operating in conjunction with the grid, with each case being separately negotiated taking into account the specific circumstances. It is also of interest to note that technical guidelines for the design of the power conditioning and protection systems for such generators have also been published in the United Kingdom by the utility (56,57).

3.3 Scope for photovoltaics in Europe

General

The commercial market for photovoltaics in Europe as elsewhere will be largely dependent on economic factors and on finance. Economic analysis will establish which of several available options is most cost effective and the availability of finance will determine how fast the favoured options will be implemented.

There are, of course, factors besides commercial preference to be considered, including the important matters of government policy and personal choice. Governments may decide that a certain proportion of power should be generated by photovoltaics in order to reduce dependency on fossil fuels and imported oil. Individuals and private companies may decide to install photovoltaic systems because they provide security against utility failure or because they operate silently and need minimal maintenance. Military authorities may decide that photovoltaic generators have significant advantages over diesel generators or batteries in applications such as mobile equipment or remote sensors.

Because photovoltaic electrical generation is a relatively new technology, unfamiliar to the present generation of engineers and administrators, it would be unreasonable to expect the rapid introduction of photovoltaic devices and systems even if the economics were favourable and the finance was available. It will take time for the new concepts and techniques to become widely understood and accepted. That is not to say that the market for photovoltaics cannot develop rapidly but rather that to do so there must be concerted efforts by all concerned for this to happen. It will be necessary to sell not only the hardware but also the concept.

The photovoltaic manufacturers have an important role here, as do the distributors of systems, in explaining the advantages and limitations of photovoltaics. They will be in competition with well-established methods of providing electricity and their first priority must be to develop and market reliable systems at competitive prices. This calls for considerable investment, not only for the installation of large-capacity automated production facilities, but also in demonstration plants and market development.

The attitude of the utilities will be of crucial importance to the widespead introduction of grid-linked photovoltaic systems. Not only will they need to accommodate the technical and administrative problems of having a large number of private 'non-firm' generators linked to their distribution system, but in time they may themselves decide to install major central generation plant made up of photovoltaic arrays thereby adding to the pool of knowledge and experience.

Governments and local civil authorities will also be involved since they will have responsibility for regulating the growing photovoltaic industry, setting standards, establishing tax and other incentives, helping manufacturing capacity to be built up, and so on. They will also have prime responsibility for resolving the inevitable conflicts of interest between competing energy sources during a period when the established pattern of electricity generation and distribution will be changing.

Small stand-alone systems

The market for small stand-alone photovoltaic systems (from a few watts to about 1000Wp) is the one least subject to objections and difficulties posed by the utilities, by governments or from other members of the public on environmental grounds. It is also the area where photovoltaics are first likely to be economic, in some applications even at today's prices.

Although it is known that most photovoltaic manufacturers have studied the market potential for stand-alone systems, their conclusions are rarely published. A recent study in France by COMES identified over 20 applications and listed the associated photovoltaic array power required (58). The applications of main interest in Europe may be divided into two categories: the consumer market and the industrial market. The consumer market consists mainly of systems that would be purchased as off-the-shelf items for remote, off-grid houses, holiday and mobile homes, such as:

- lighting units (20-100Wp)

- small pumps (200-500Wp)

- refrigerators of various types and sizes (200-600Wp)

- television sets (50-100Wp)

- battery chargers of various types and sizes (50-250Wp)

- security devices, such as intruder alarms, electric fences, fire alarms, etc (10-50Wp)

The industrial market for small stand-alone systems will also consist largely of packaged systems for remote locations. Again these would be mainly purchased as off-the-shelf items and typical applications include:

- marine beacons, navigation lights and fog signals (50-250Wp)

- lighting units for mountain refuges and small buildings (100-250Wp)

- aircraft beacons on tall buildings and in remote locations (50-200Wp)

- automatic weather stations and other remote instrumentation stations such as river level gauges (100-250Wp)

- portable radio-telephones and highway emergency telephones (20-50Wp)

- railway signalling (50-150Wp)

- small radio and television transmitters (500-1000Wp)

- small radar stations (500-1000Wp)

- fans and other small devices for agricultural applications (200-1000Wp)

- highway warning signs (100-250Wp)

- cathodic protection systems for wellheads, pipelines, bridges and other steel structures (50-500Wp)

- integrated security systems (10-50Wp)

Examples of most of these small stand-alone systems are available commercially today, although in many cases further development remains desirable, to optimise the system as a whole. Many installations based on full life-cycle costs and at current system prices of ECU 20-35/Wp are already competitive with alternative power sources. A steady increase in sales may be anticipated as potential customers become informed of the availability of such systems and they hear favourable reports from satisfied users.

There is no precise method of forecasting sales of these relatively small stand-alone photovoltaic systems, but given large scale production and effective marketing, it seems probable that by 1990-1995, complete systems will be available for less than ECU 5/Wp. Considering the whole of western Europe with a total population of over 300 million, a large holiday industry and with many thousands of off-grid houses, particularly in southern Europe, the following mature market projection would seem reasonable:

a) Consumer systems of all types - 100000 units per year by 1995. The average size of the photovoltaic array for each system is likely to be about 50Wp and thus the total power involved would be about 5MWp/year.

b) Industrial systems of all types - 50000 units per year by 1995. The average size of the photovoltaic array for each system is likely to be about 150Wp and thus the total power involved would be about 7.5MWp/year.

It is anticipated that this level of market activity for small stand-alone photovoltaic systems amounting to about 12MWp/year, would not increase significantly after 1995, even if system costs continued to decline, since the market in Europe will have become fully developed. The export market however will probably continue to grow steadily and is expected to exceed by far the European market.

The projected development of the market for small stand-alone systems in Europe in shown in Figure 3.23.

Larger stand-alone systems

The market for larger stand-alone systems in the size range 1 to 300kWp is mainly that at present occupied by batteries or gasoline and diesel generators. Batteries or another form of energy storage will generally be required with these systems to provide continuous power capability. This has a significant effect on the economics of such systems and also on the applications where they could be used. In general they would not be suitable for use as portable generators due to the size of arrays and the great weight of batteries involved. The low use factors typical of this type of generator application will result in a high cost of power. There are however several situations where stationary generators operating at high use factor are used, providing power for such applications as:

- off-grid houses and villages (2-300kWp)

- large pumps, in which case battery storage may in general be avoided (1-100kWp)

- radio and television transmitters and micro-wave repeater stations (1-100kWp)

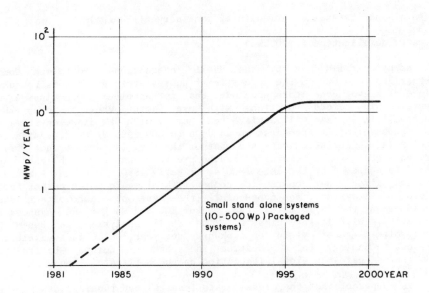

Fig 3.23   Market projection for small stand-alone
photovoltaic systems in Europe

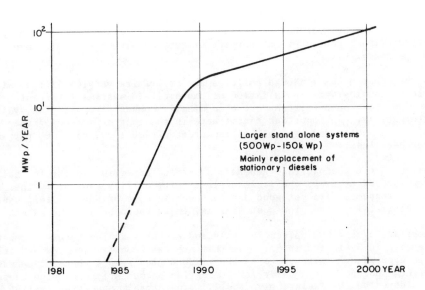

Fig 3.24   Market projection for larger stand-alone
photovoltaic systems in Europe

- radar stations and similar (10-150kWp)

- rural industries and agricultural machinery (1-300kWp)

- water desalination (5-100kWp)

A study carried out by the Shell International Petroleum Company, Netherlands and UK, on the economics of photovoltaic generators to take the place of diesel generators concluded that for stationary continuous power systems in remote locations with use factor greater than 20-25%, photovoltaic systems will achieve cost parity with 5kW diesel systems when the photovoltaic system cost comes down to ECU 6 to 9/Wp, depending on the fuel relative inflation rate and whether the fuel is taxed or not (59). On present forecasts, as discussed in the previous chapter, such cost levels could be achieved by the late 1980's, or early 1990's. Larger sized diesel systems give lower costs due to benefits of scale which do not apply to photovoltaic systems, and consequently the cost of photovoltaic systems would need to fall below the range ECU 6-9/Wp for such systems to be competitive with larger diesels. The Shell analysis was based on a photovoltaic power yield of 1.6kWh/Wp per year, which is equivalent with present technology to a location in southern Europe having about 1750kWh/m^2 per year global insolation on the horizontal plane.

It is apparent that the diesel costs (capital and running) assumed in the Shell study relate to remote European or similar conditions. In many developing countries, diesel costs are considerably higher and hence the break-even photovoltaic costs would be higher than the ECU 6 to 9/Wp referred to above. Such applications are close to being economic today.

It must be noted however that for private purchasers, photovoltaic systems must be somewhat cheaper than the above break-even costs with diesel generators for significant markets to open up. This is because most of the potential private purchasers of such systems are relatively poor and cannot afford diesel generators at current prices, or else they would have bought them already.

On the other hand, although policies and priorities vary, utilities will be able to consider the installation of photovoltaic systems for remote parts as soon as such systems are economically competitive with the available alternatives. An important factor here is the cost of extending the grid, which is estimated to be of the order of ECU 10000 to 17000 per km for overhead lines.

The scope for photovoltaic generators serving off-grid housing in Italy has been studied by ENEL, the Italian electricity utility (60, 61). The short term prospects are not good due to the high cost of photovoltaic systems, but ultimately, given low cost systems, there could be a large market.

There are over 70000 houses in Italy and probably at least this number in Greece, in Spain and in Portugal that are permanently occupied and are not grid connected. There are a further 20000 in France but elsewhere in western Europe the number of off-grid permanently occupied houses is considered to be relatively small. The total potential market for photovoltaic systems for off-grid houses could thus be of the order of 300000. System sizes would range from about 1.5kWp to over 7kWp, with an average of about 2.5kWp (ie giving about 2.5 to 3.0kWh/day during the winter months in southern Europe). Based on the breakeven conditions indicated on Figure 2.31, given low cost systems (ie, less than about ECU 3/Wp) and subject to appropriate finance being forthcoming, significant penetration could start in 1985 and rise to about 3000 to 4000 systems a year by 1990 and level off at about 6000 to 8000 systems a year by 1995. The associated photovoltaic array market would thus be about 10MWp per year in 1990, rising to about 20MWp per year in 1995, with little increase thereafter likely. In time this market will decline, as all the off-grid houses are equipped with generating systems of one type or another. Costs will need to fall further before stand-alone photovoltaic systems would be economic for off-grid houses in more northerly regions of Europe but because of the relatively small numbers this will not significantly affect the total market.

A similar size of market may be expected to develop for off-grid houses that are not permanently occupied (ie, mainly holiday and weekend houses). In view of the higher spending power of the people who use these houses and their interest in having a silent and virtually maintenance-free electricity supply, photovoltaic systems are likely to prove an attractive alternative to diesel generators, even if not strictly economically cheaper. There are well over 500000 such houses in the whole of Europe. Given reliable, well-engineered and marketed systems, significant sales could start within a few years, reaching 4000 to 5000 systems a year by 1990, growing to perhaps 8000 to 10000 system a year by 1995 and remaining at around that level thereafter. The average size of system would probably be about 2kWp, somewhat less than for permanently occupied houses, even though the lifestyle of the occupants would call for higher electricity usage (eg, many labour saving devices), since holiday houses are mainly occupied only in the Summer season, with associated higher insolation. The photovoltaic market would thus be about 10MWp per year in 1990, rising to about 20MWp per year in 1995 and thereafter.

One other important application that will become economic at lower system costs is stand-alone large systems to supply remote villages and island communities which at present either are served by diesel generators or have no electricity supplies. There are probably at least 500 such locations in southern Europe, for which the size of the photovoltaic system would need to be between 50 to 150kWp. Assuming some 50 of these per year were to be equipped with photovoltaic generators throughout the 1990's, the market would be about 5MWp per year.

The projected market for large stand-alone systems in Europe is shown in Figure 3.24.

Grid-connected residential systems

The potential for grid-connected residential photovoltaic systems in Europe has not been studied in depth although organisations in France and Italy have carried out some work in this area (58, 62). In the USA such systems are considered by many to offer an early potential market and one which will grow rapidly. The commercial readiness price goals for residential systems, as set by the US Department of Energy Photovoltaics Program, were as follows (1980 dollars, approximately equivalent to ECU):

PV Module price (FOB): $0.70/Wp
System prices: $1.60-2.20/Wp
Production scale: 100-1000MW/year
User energy price: $0.05-0.09/kWh

The system price correlates with the production scale and the user energy
price range reflects variations in location (ie, solar insolation), system
price and utility buyback arrangements. The original expectation was that
the module and system price goals would be achieved by 1986, but it has to
be recognised that the photovoltaic program in USA is currently undergoing
some revision due to financial cut-backs and it is now considered that the
price goals will consequently not be achieved until several years later.
No major technical obstacles are however foreseen that would prevent the
goals being attained in due course.

A major point in favour of residential photovoltaic systems is that the
load is close to the generation source and as a result there are minimal
transmission and distribution losses. From the user's viewpoint, the
reliability of his supply is improved, particularly if battery storage is
incorporated. From the utilities viewpoint, the photovoltaic generator
contributes to the total generating capacity of the system, thereby
reducing the capacity that has to be provided by conventional generating
plant.

If installed as part of a new building during construction, the
photovoltaic array may be incorporated into the roof structure and may,
with some designs, replace part of the conventional waterproofing system,
giving appreciable cost savings. Retrofit installations (ie, installing
the array on existing houses) would be considerably more expensive.

Being connected to the grid, the need for a large amount of expensive
energy storage in the form of batteries is avoided, since surplus energy
would be fed to the grid during sunny periods and energy would be drawn
from the grid at night and during cloudy periods. A certain amount of
battery storage would nevertheless generally be desirable to smooth out
fluctuations and, depending on the utility rate structure, enable energy to
be taken from the grid only at times when lower tariffs were applicable.

Extensive technical, economic and market studies would be needed to
determine the potential for grid connected residential photovoltaic systems
in each European country. For present purposes, an indication of the
prospects can be gained from the break-even curves presented in
Figure 2.31. There would seem to be no unsurmountable technical barriers
to prevent photovoltaic systems coming down in price levels that would
enable them to generate electricity at costs comparable with grid selling
prices in southern Europe by about 1990 and in northern Europe by about
2000, given continued real inflation in the price of grid-supplied
electricity. It would appear quite feasible, given the necessary finance
and political will, for most if not all new single family houses and a fair
proportion of low-rise apartments within the break-even areas to be
equipped with photovoltaic systems. A few years after systems for new
houses started to be cost-effective, retrofit installations would also
become cost effective, assuming continued real price inflation of grid
supplied electricity plus possibly some further reduction in system prices.

The market for photovoltaic systems would thus be dependent, at least initially, on the new house building rates in each country. This of course is extremely difficult to predict, dependent as it is largely on the general economic climate. As a first approximation, an average rate of four new dwellings per 1000 population per year is considered a conservative assumption. Based on a total southern European population of about 100m in 1990, and assuming at that time 10% of the new houses are equipped with photovoltaic systems of about 5kWp each, the corresponding market would be 200MWp/year. Central Europe, with a comparable total population, could provide a similar market by 1995 and northern Europe, again with a total population of about 100m, would provide a further 200MWp/year by 2000. A further 200MWp/year may be allowed for retrofit installations, which similarly would start in southern Europe and work north, giving a total European market of about 800MWp/year by the year 2000, with the market continuing to grow rapidly.

Further strong growth of the market for grid connected residential systems may be expected as the proportion of new houses equipped with photovoltaic generators steadily increases. The retrofit market would also continue to grow and in time there may arise a market for systems to replace earlier installations. It is thus conceivable that the total market in Europe for grid connected residential systems could grow to over 2000MWp/year by the year 2025.

It is of interest to compare these figures for Europe with the findings of studies carried out in the USA to estimate the potential market there for such systems (for example, reference 63). Figure 3.25 shows an estimate of the annual demand in the USA related to the module cost. (The system cost would be some $0.90–1.50/Wp greater than the module cost). For a module cost of $0.70/Wp, the potential market is estimated to be about 600MWp per year, rising to over 2000MWp/year if module costs came down to about $0.30/Wp. After allowing for the differences in population, these market estimates are somewhat higher than the European market projection, which is as might be expected given the more favourable solar regime in the USA coupled with a more affluent and generally more technically adventurous society.

The projected market for grid-connected residential systems in Europe is shown in Figure 3.26.

Photovoltaic systems for the service, commercial,
institutional and industrial sectors.

The service, commercial, institutional and industrial (SCII) sectors of the economy encompass many business establishments with electrical loads in the range 25kW to 5MW. Many of the 16 European Community photovoltaic pilot plants come into this category. These loads are intermediate between those of residences, which are generally considerably less than 25kW, and those of conventional central electricity generating stations, whose capacities range from 100MW to over 1000MW.

Again detailed studies on the potential for photovoltaic systems for the SCII sectors have not been carried out for Europe and for present purposes only a preliminary assessment is possible. The first point to note is that there is a tremendous variety of possible applications leading to a great diversity of operating schedules, energy demands, space availability and

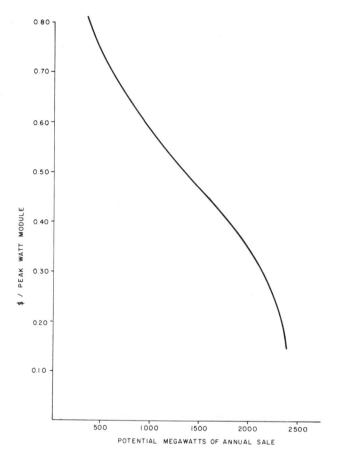

Fig 3.25 <u>Total annual potential demand for residential</u>
<u>photovoltaic power systems in USA</u>

access to solar energy.

It is also important to note that, for many electricity users in the SCII sectors, the consequence of failure in the grid electricity supply are very serious, and expensive stand-by diesel plant has to be maintained to meet essential loads in the event of power failure. Photovoltaic systems with energy storage could find a practical application as a second source of power, used to supplement grid power normally and meet all emergency demands when necessary long before such systems become directly competitive with utility supplied electricity.

The electrical energy consumption for the top 20 SCII sub-sectors in the USA is shown in Table 3.8. The consumption distribution is important since the combined capacity of many smaller users in the non-industrial sectors is shown to be commensurate with industrial sectors. These smaller users then represent the potential for demonstration of small individual systems requiring less investment risk, but which represent a large potential photovoltaic market. A similar situation is likely to obtain in Europe.

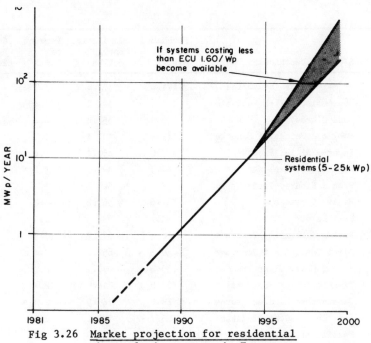

Fig 3.26 Market projection for residential
          photovoltaic systems in Europe

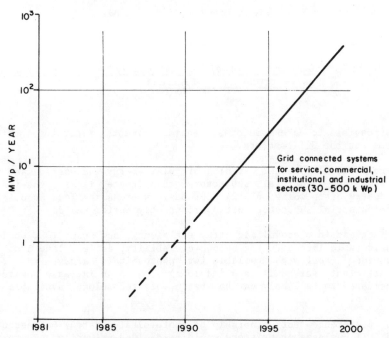

Fig 3.27 Market projection for service, commercial,
          institutional and industrial photovoltaic
          systems in Europe

Order	SIC Code Description	Purchased Electric Energy TWh	Cumulative Energy TWh	Percent of Total	Cumulative Percent of Total
1	Primary Metal Industries	171.8	171.8	13.7	13.7
2	Chemicals, Allied Products	130.6	302.4	10.4	24.0
3	Food Stores	55.4	357.8	4.4	28.5
4	Public & Parochial Schools	49.6	407.4	4.0	32.5
5	Paper & Allied Products	43.0	450.4	3.4	35.9
6	Pipe Lines, excl Nat Gas	40.5	490.9	3.2	39.1
7	Food & Kindred Products	38.8	529.7	3.1	42.2
8	Health Services	37.6	567.2	3.0	45.2
9	Real Estate	31.1	598.3	2.5	47.7
10	Eating & Drinking Places	30.9	629.2	2.5	50.1
11	Elect Gas & Sanit Services	30.5	659.7	2.4	52.5
12	Stone, Clay, Glass Products	30.4	690.1	2.4	55.0
13	Transportation Equipment	29.8	719.9	2.4	57.3
14	General Merchandise Stores	29.6	749.5	2.4	60.0
15	Petroleum & Coal Products	28.7	778.2	2.3	62.0
16	Textile Mill Products	28.3	806.5	2.3	64.2
17	Machinery, Except Electric	27.4	833.9	2.2	66.4
18	Fabricated Metal Products	26.5	860.4	2.1	68.5
19	Electric, Electronic Equipment	25.9	886.3	2.1	70.6
20	Nonclassifiable Establishments	20.2	906.5	1.6	72.2

Table 3.8   Consumption of electrical energy by the top 20 SCII users in USA (1974 figures)

It is possible to make some other general comments regarding photovoltaic systems for the SCII sectors:

- There is usually better matching of solar energy and user demand profiles in the non-industrial subsectors than there is in the industrial. (Industry often works on a 24 hour basis whereas the load associated with non-industrial subsectors often arises only during the day).

- An increase in photovoltaic array efficiency, now about 10%, would have a considerable impact on the economic viability, as less land area would be occupied. (Roof area available for photovoltaic arrays is frequently insufficient for SCII applications). Such an increase however is not considered to be likely on the basis of technology available now or foreseen.

- The preference for ownership of photovoltaic systems between utilities and SCII business proprietors would be decided primarily by tax status and the financial terms applicable. Non-taxed institutions (schools, hospitals, etc.) may have certain advantages in this respect over tax-paying businesses, depending on the way capital and operating costs are financed and taxed.

The economic viability of photovoltaic systems in the SCII sectors before the year 2000 is strongly dependent on the assumption that electricity prices will escalate at a rate at least 5% higher than the general inflation rate and that complete systems become available at prices less than about ECU 3/Wp. As discussed at the end of the previous chapter, there are good grounds for expecting that these conditions will be met with either ribbon silicon or thin film techniques.

A study carried out in 1979 in the USA concluded that there was a good potential for photovoltaic systems serving the SCII sectors in the USA (64). The photovoltaic potential is presented in Table 3.9. It should be noted that the units used to indicate photovoltaic potential in this table give a relative measure only of the potential displacement of energy demand from other sources taking into account only technical factors such as total energy use in the subsector and insolation levels.

When economic factors such as installed costs, electricity rates, operation and maintenance costs are included in the analysis, the resulting application rankings become as shown in Table 3.10. The Energy Market Potential (EMP) indicated on this table is a rating factor determined by (i) the electrical consumption and (ii) the expected profitability of the photovoltaic system, that is, the after-tax benefit to the owner over the life of the photovoltaic system expressed as a fraction of the initial investment.

From this it was concluded that the annual total electrical energy that could be produced by potential photovoltaic systems in the SCII sectors in the USA would amount to 680×10^9 kWh. This would imply the installation of some 500000MWp of photovoltaic systems, implying a rate of penetration that almost certainly could not be achieved by the year 2000 even if economically viable systems started to become available in 1990. The most that could reasonably be expected is about 5000MW by the year 2000.

Obviously great caution must be exercised when attempting to interpret and use such figures, derived as they are on the basis of so many assumptions. However, in relation to Europe, it is probable that similar economic arguments will apply for the SCII sectors, starting first in the southern European countries and progressively working north as systems became cheaper and grid electricity more expensive. An important difference is that the rate of penetration may be expected to be somewhat slower in Europe than in the USA due to greater financial constraints and a generally more conservative approach to new technology. The total installed capacity of photovoltaic systems in the SCII sectors in Europe would on this basis be unlikely to exceed 1000MWp by the year 2000, assuming economic viability is achieved by 1990. A possible development would be (excluding subsidised demonstration plants) 10 such installations in 1990, 20 in 1991, 40 in 1992 and so on, the number doubling each year. This would result in a cumulative total of about 10000 installations by the year 2000, requiring about 1000MWp of photovoltaics, on the assumption that the average size of an SCII installation will be 100kWp. The rate of installation in the year 2000 would be about 4000 systems per year, say 400MWp of photovoltaics. This scale of business would continue to grow until well into the next century.

Order	SIC Code	Short Title	Photovoltaic Potential		Order	SIC Code	Short Title	Photovoltaic Potential
1	28	Chemicals, Allied Products	182616		34	70	Hotels, Oth. Lodging Places	15902
2	33	Primary Metal Industries	162138		35	56	Apparel & Accessory Stores	12114
3	54	Food Stores	77426		36	14	Nonmetallic Min. Except Fuels	11058
4	82	Public & Parochial Schools	65177		37	60	Banking	10782
5	80	Health Services	65168		38	75	Auto Repair, Services, Garages	9708
6	32	Stone, Clay, Glass Products	54964		39	57	Furniture, Home Furnish. Stores	8246
7	53	General Merchandise Stores	62368		40	23	Apparel, Oth. Textile Prods.	7134
8	20	Food & Kindred Products	51999		41	52	Build. Materials, Hdware, etc.	7057
9	58	Eating & Drinking Places	51517		42	38	Instruments, Related Prods.	6732
10	46	Pipe Lines, Exc. Nat. Gas	50819		43	25	Furniture & Fixtures	6223
11	26	Paper & Allied Products	49506		44	63	Insurance	5650
12	49	Elect. Gas, & Sanitary Services	44264		45	27	Printing & Publishing	5094
13	65	Real Estate	41065		46	39	Misc. Manufacturing Indus.	4625
14	35	Machinery, Except Electric	40695		47	61	Credit Agencies, Non-Bank	4547
15	29	Petroleum & Coal Products	37526		48	45	Air Transportation	4531
16	22	Textile Mill Products	34917		49	79	Amuse. & Recreation, Non-Film	4040
17	34	Fabricated Metal Products	33283		50	76	Misc. Repair Services	2983
18	36	Electric, Electronic Equip.	31712		51	64	Ins. Agents, Brokers, & Services	2659
19	42	Motor Freight Trans. & Wareh'ing	29604		52	81	Legal Services	1818
20	50	Wholesale - Durable Goods	28886		53	40	Railroad Transportation	1662
21	37	Transportation Equipment	28110		54	78	Motion Pictures	1478
22	99	Nonclassifiable Establishments	27361		55	21	Tobacco Products	1422
23	55	Auto Dealers & Services Sta.	25330		56	31	Leather, Leather Products	1044
24	86	Membership Organizations	25168		57	84	Museums, Galleries, Gardens	826
25	30	Rubber, Misc. Plastics Prod.	23163		58	91	Gen. Gov.; Non-Finance	584
26	73	Business Services	22736		59	41	Local & Highway Passenger Trans.	568
27	13	Oil & Gas Extraction	22582		60	62	Sec. & Com. Brokers, Exchanges	424
28	89	Misc. Services	22129		61	93	Pub. Finance, Tax & Mon. Policy	321
29	59	Misc. Trade	20778		62	44	Water Transportation	169
30	83	Educ. Services, Other	20740		63	92	Justice, Pub. Order, Safety	164
31	48	Communication	20153		64	67	Holding, Oth. Invest. Offices	130
32	24	Lumber & Wood Products	17069		65	51	Wholesale - Nondurable	113
33	72	Personal Services	16647					

Table 3.9 Photovoltaic systems potential for the SCII sectors in USA

Rank	SIC	Short Title	Energy Market Potential $EMP \times 10^{-9}$	Expected Profitability EP	Annual Electric Energy $E(TWh)$
1	86	Membership Organisations	11.4	.611	16.2
2	80	Health Services	11.3	.264	37.3
3	82	Public & Parochial Schools	11.0	.186	49.2
4	54	Food Stores	9.0	.141	54.9
5	58	Eating & Drinking Places	7.6	.213	30.6
6	83	Educational Services, Other	6.4	.394	15.5
7	53	General Merchandise Stores	4.6	.136	29.4
8	73	Business Services	4.6	.199	18.4
9	35	Machinery, Except Electric	4.2	.167	27.0
10	50	Wholesale - Durable Goods	3.7	.161	19.5
11	59	Misc. Trade	3.3	.255	11.4
12	89	Misc. Services	3.3	.167	15.6
13	70	Hotels, Other Lodging Places	3.1	.153	17.9
14	42	Motor Freight Trans. & Wareh'ing	3.1	.136	18.8
15	34	Fabricated Metal Products	2.8	.106	26.1
16	72	Personal Services	2.5	.241	9.2
17	49	Electricity, Gas, & Sanitation Ser.	2.5	.064	30.2
18	32	Stone, Clay, Glass Products	2.3	.078	30.0
19	65	Real Estate	2.1	.039	30.8
20	56	Apparel & Accessory Stores	1.9	.266	6.4
21	36	Electric, Electronic Equipment	1.9	.062	24.8
22	75	Auto Repair, Services, Garages	1.6	.236	5.7
23	20	Food & Kindred Products	1.5	.044	38.3
24	48	Communication	1.5	.097	13.1
25	60	Banking	1.4	.128	9.0
26	57	Furniture, Home Furnish. Stores	1.1	.233	4.4
27	37	Transportation Equipment	1.1	.073	29.3
28	52	Building Materials, Hardware, etc.	0.9	.251	3.5
29	38	Instruments, Related Products	0.9	.153	4.4
30	61	Credit Agencies, Nonbank	0.9	.243	3.0
31	63	Insurance	0.8	.181	3.5
32	55	Auto Dealers & Service Stations	0.7	.025	16.6
33	39	Misc. Manufacturing Industries	0.6	.105	4.0
34	23	Apparel, Other Textile Products	0.6	.077	6.5
35	76	Misc. Repair Services	0.6	.250	2.0
36	25	Furniture & Fixtures	0.6	.153	4.2
37	45	Air Transportation	0.5	.125	3.3
38	81	Legal Services	0.4	.314	1.0
39	78	Motion Pictures	0.3	.167	1.3
40	27	Printing & Publishing	0.2	.012	9.0

$$\text{Total Area Related Cost} = \$100/m^2$$
$$\text{Inverter Cost} = \$100/kW$$
$$\text{Battery Cost} = \$40/kWh$$

Table 3.10 Application ranking for low-cost photovoltaic systems for SCII sectors in USA

The projected market for the SCII sector in Europe is shown in Figure 3.27.

Photovoltaic central generating stations

The rate at which photovoltaic systems are integrated into the electricity utility generating mix will depend upon several factors and not simply on the economics. National policies may require the utilities to provide a certain proportion of generating capacity from indigeneous sources to reduce dependence on imported fuels. The utility may find that a photovoltaic plant, being incremental, can be built up at a rate to suit a rising demand, whereas a major thermal installation would take much longer to bring on stream. Photovoltaic systems may require less skilled operating and maintenance staff, an important consideration in remote areas. On the other hand, limitations on land use may restrict the introduction of the large arrays associated with central generating plants.

There is nevertheless no doubt that economic criteria will play the dominant part in determining the role for photovoltaic systems for central generation. The key factors will be (i) the cost savings that will accrue from displacement of fossil fuels whose prices will rise faster than general inflation; and (ii) the displacement of a significant proportion of conventional generating capacity (capacity credit).

There have been few studies in Europe on the potential for photovoltaic central generators, although there are plans for a 1MWp experimental plant in Italy, the DELPHOS project (65). Some consideration has been given to the feasibility of building ground rectennas in Europe for large photovoltaic space power stations (66), but for various reasons this approach is not considered appropriate, at least for Europe.

In the USA, the following price goals were established for photovoltaic central generators (in 1980 dollars):

PV module price (FOB):	$0.15-0.40/Wp
System price:	$1.10-1.80/Wp
Production scale:	500-2500MWp/year
Unit energy price:	$0.04-0.08/kWh

Whether these very low price goals can ever be achieved is unclear at this time, since they are dependent on the successful development of low cost thin film photovoltaics, which are unlikely to be proven for widespread commercial applications until the early 1990's. In any case, it would appear that higher system prices could be acceptable on economic grounds, giving continued real price inflation of conventional or nuclear alternatives, as was discussed at the end of Chapter 2.

The key findings of various studies in the USA and in Europe on photovoltaic central generators may be summarised as follows:

- Established utility generation planning methods are valid for studying photovoltaic generation with minor modifications to allow for the special features of photovoltaic systems.

- Photovoltaic plants, particularly if dispersed over a wide geographic area, do have certain capacity value: ie, they can be considered as contributing to the effective installed capacity and not serve just as fuel-savers. Such systems would contribute to meeting the peak system loads with a given level of reliability. The potential capacity credit however has been derived using probability and statistical methods and practical acceptance will require further investigation and operating experience.

- The value to the utility of both the effective capacity and the energy output of a photovoltaic plant depend strongly on the characteristics of the utility system to which they are applied. The cost and types of conventional generation and the relative times of daily load and solar radiation peaks are particularly important.

- As photovoltaic plant penetration increases, the incremental value of the energy supplied decreases, since the less efficient generating units will have been displaced first. Subsequently displaced units have lower energy costs and hence the value of the displaced capacity diminishes.

- Energy storage dedicated to the photovoltaic plant offers limited advantages. Energy storage has its greatest value as system storage, designed and operated for the benefit of the total network. If storage is sufficiently cheap to be added to a photovoltaic central generation station to provide the required reliability and thus displace generation capacity elsewhere, the same storage without the photovoltaic system would also allow capacity to be displaced. For example, storage could be charged during off-peak time by low-cost base load plant, irrespective of whether the system included photovoltaic generators.

The overall conversion efficiency of photovoltaic systems based of crystalline silicon cells is currently about $5.5 - 7.5\%$. Technical developments may be anticipated that will improve the overall efficiency to $8 - 10\%$. Systems based on thin film cells will probably have somewhat lower overall efficiencies, of the order of $5 - 7\%$. Thus, land areas of between 1.0 and $1.5km^2$ will be needed for each 100MWp of central generating plant. In the USA there is in generally plenty of available land, particularly desert areas, where large photovoltaic arrays could be built. For Europe however, the availability of land is likely to prove an important consideration for large photovoltaic generators, since competing interests are involved. The siting of large photovoltaic plants will need to be considered as part of national energy and land use policies, taking into account relevant factors including costs and amenity considerations. Although some local problems may be anticipated, there is in general ample land available in Europe for photovoltaic central generating plants, particularly when it is considered that near most towns there are usually considerable areas of waste land, former factory sites, disused railway lines, obsolete thermal power stations and the like which could be used without too much difficulty.

When photovoltaic systems become available at costs which make them attractive to utilities for central generation, bearing in mind strategic and environmental considerations, then we can expect to see photovoltaic power plants being built to feed into the European grid system. Or one possible development could be for many existing power stations to have photovoltaic generators built alongside, often within the same site boundary, to reduce the associated costs for land, switchgear, transmission and control.

In time, the utilities may be expected to establish numerous photovoltaic generating stations near towns wherever possible, using the waste or under-utilised land areas referred to above. The size of the sites available would not in general exceed 4 or 5km^2 limiting the size of the photovoltaic generator to about 250MWp, but the total contribution could be significant given many hundreds of such installations in each country, particularly in southern Europe.

The introduction of commercial photovoltaic central generating stations may possibly start in the late 1990's, but the main development in this important sector is unlikely to arise until later, in the first quarter of the next century. The potential size of this market is of course very large indeed and by the year 2025, the total photovoltaic capacity in Europe for central generation could be of the order of 50000MWp.

There is perhaps one other long-term possibility that should be considered and that is to build very large photovoltaic plants in the northern Sahara desert and transmit the power by very high tension submarine cables across the Mediterranean to Sicily. The power would then feed into the Italian grid, which in turn could feed the rest of the European grid. The technical problems would be formidable and the costs of transmission great, but if very low cost (less than ECU 1.50/Wp) photovoltaic systems were available, the scheme might be viable. Maintaining security of the system in a remote area given to political instability would remain a problem.

As was mentioned in Chapter 2, Dahlberg has proposed a similar concept, but instead of transmitting electricity over great distances, hydrogen-generating photovoltaic plants are postulated which would be situated in desert areas (32). This scheme is based on the premise that in the long term hydrogen will become the most appropriate substitute for oil.

Total European market for photovoltaics

The above discussions on potential markets for photovoltaics in Europe are highly speculative and depend on many assumptions, particularly regarding future total system costs. Unless there is suitable support for the emerging photovoltaic industry, it is unlikely that the low system cost targets will be achieved within the time periods mentioned, in which case most of the markets for photovoltaics in Europe will not become economically viable until much later, when conventional energy prices have further risen.

Assuming however that full support is given to the industry and that appropriate financial and fiscal encouragements are provided to purchasers of systems, then the market projections for the various categories of system may be taken as an indication of the potential. The aggregated total sales projection for photovoltaic applications in Europe to the year 2000 is shown in Figure 3.28, which shows a steady increase in annual sales (including commercial and pilot/demonstration sales) throughout the period under review. In 1990, the market could be about 20MWp per year, rising to between 650 and 1200MWp per year by the year 2000, implying average growth rates from a base of about 1MWp in 1983 of between 45% and 50% per year. A growth rate higher than 50%, although not impossible in the late 1990's would be difficult to sustain, for financial and institutional reasons. A lower growth rate, associated with the lower total sales line included on Figure 3.28 may obtain if grid-connected residential and SCII systems do not become economically viable within the time frame being considered.

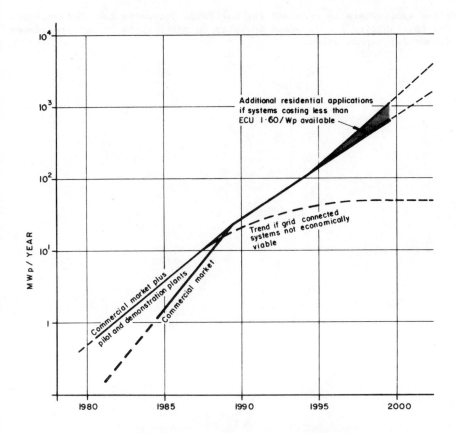

Fig 3.28 <u>Aggregate total market projection for</u>
<u>photovoltaic systems in Europe</u>

Assuming the low cost targets are achieved as expected, then the photovoltaic market in Europe will continue to expand vigorously after the year 2000 for at least another 10 to 15 years before the classic S-curve for the penetration of a new technology begins to curve over as saturation is approached. By the year 2025, it is quite conceivable that the total installed capacity of photovoltaic systems in Europe could exceed 200GWp, generating about 10% of the total electricity supplied.

The total value of the sales of photovoltaic systems for applications in Europe (ie, excluding export sales) can be derived using typical system costs. Assuming the cost of systems falls in the range indicated on Figure 2.27, the total value of photovoltaic system sales for applications within Europe would be:

```
1983     ECU     10 -    15 million per year
1990     ECU     80 -   100 million per year
2000     ECU   1500 -  2000 million per year
```

It is now appropriate to consider the worldwide prospects for photovoltaics and the oportunities for European industry to participate in aid programmes and commercial export sales.

CHAPTER FOUR - PHOTOVOLTAIC PROSPECTS WORLDWIDE

4.1 The international scene

Co-operation in photovoltaics

Photovoltaic solar energy is widely recognised as being of great practical significance for many developing countries. The biggest immediate potential for photovoltaic systems lies in the provision of electricity for remote villages and small stand-alone systems for water pumping, refrigeration, ice-making and desalination. Many governments of developing countries, recognising the need to build-up experience with such systems, have made arrangements with international funding agencies, industrialized countries and individual manufacturers for various solar energy projects including photovoltaics. The European Community has provided funds for several photovoltaic projects in developing countries including:

- Cameroun: 5kW photovoltaic solar pumping system for an irrigation project

- Comoros: photovoltaic generator for electricity supply for a herzian relay

- Egypt: assistance in setting-up EREDO, a centre for the application of renewable energy sources

- Jordan: scientific co-operation for solar energy applications with the Royal Scientific Society

- Mali: photovoltaic pumping systems at seven locations for drinking water and irrigation supplies

- Senegal: photovoltaic pumping systems at ten locations for drinking water and irrigation supplies

- Syria: scientific co-operation for solar energy applications with the CERS

Several other CEC-funded projects for the provision of photovoltaic pumping systems are being planned (see Appendix C).

A number of developing countries are establishing their own capabilities in the field of manufacture of photovoltaic devices and systems. India for example has a self-sufficient photovoltaic programme and has already built a pilot plant for making photovoltaic cells and modules. A number of systems have been developed and are being demonstrated at various sites in India. Organisations in Mexico, Brazil and Singapore are also developing their own manufacturing capability for photovoltaics. In Africa, Photowatt International, a French manufacturer of photovoltaic systems, is establishing a manufacturing facility at Abidjan, Ivory Coast, in response to the rapidly growing market there for photovoltaic systems, particularly for educational television sets.

The largest international co-operation programme in the solar energy field at present is between Saudi Arabia and the USA. The Soleras project, with a budget of more than US $20 million, includes the construction of a 350kWp photovoltaic generator to supply two villages with electricity, with provision for performance evaluation and training support (Figure 4.1). Unfortunately, the high-technology concentrator modules being used for this project are likely to prove difficult to operate and maintain, particularly in view of the site conditions.

Fig 4.1   350 kWp concentrator photovoltaic system for three villages in Saudi Arabia, part of SOLERAS project (Source: Martin Marietta Aerospace Solar Energy Systems)

The most successful international projects are those which incorporate technology truly appropriate to the circumstances and which offer good possibilities for local participation, leading in time to local manufacture of systems. International links between companies engaged in the manufacturer and distribution of photovoltaic systems are discussed in Chapter 5.

Energy problems in developing countries

Energy consumption in developing countries has been growing much faster than for the world as a whole, but the total demand is still relatively small. In 1980, commercial energy consumption in the developing countries amounted to 830 mtoe or about 12% of the total world commercial energy consumption (excluding China). The per capita consumption is consequently very low.

Most concern focuses on the oil-importing developing countries, the majority of which import 75 to 100% of their commercial energy requirement as oil. Seventy-three countries fall into this category and all but 24 of them also have serious fuelwood problems, according to World Bank statistics. The 73 highly vulnerable countries account for some 400m people. Over 1600m people, less then half the world's population, live in developing countries currently obliged to import large quantities of oil.

Some developing countries, such as Indonesia, Nigeria and China, which at present are net oil exporters, may cease to be so within the next decade, as their internal consumption rises with development. Although many low income countries (ie, countries with per capita GNP less than US $360 per year by the World Bank definition) derive over half of the total energy from wood and agricultural or animal wastes, their ever-growing cities depend heavily on oil. The energy problems, already acute, will get worse unless these developing countries are helped to plan, organise and invest in indigeneous energy resources.

Table 4.1 shows the principal investment requirements in commercial energy 1980-1990 for the oil-importing developing countries, as forecast by the World Bank. By 1985 the investment requirements for commercial energy in the oil-importing developing countries are expected to reach US $50 billion per year, of which only $0.1 billion is expected to be devoted to renewables for electricity production.

By 1990, the total investment requirements are forecast by the World Bank to rise to $60 billion per year, with a relatively much greater proportion being devoted to renewables for electricity production, estimated to be of the order of $0.7 billion per year. The major part of this investment could be devoted to the introduction of distributed photovoltaic generating systems serving mainly the rural areas, since by that date such systems are likely to be technically and economically competitive with alternative generation systems, as discussed at the end of Chapter 2.

4.2 Photovoltaic systems for developing countries

The market

The potential market for photovoltaic systems in the developing countries is immense. It has been estimated that there are some 10 million villages in these countries, with populations of 100 to 1000 persons each. Nearly half the world's population live in these villages and they all need power for pumping water for drinking and irrigation purposes, for food processing and refrigeration, for powering small workshop equipment, for lighting, for village dispensaries, for telecommunications and a host of other purposes. Electric power for these basic needs will enable their living standards to improve, raise agricultural and village industry productivity and help provide food for the world's growing population. Photovoltaic generators provide an appropriate means of meeting this great need, need, since they are simple to operate and maintain, require no fuel and will last many years. Moreover, most of the developing countries concerned have high levels of solar insolation and usually there is plenty of land available for the photovoltaic arrays.

To obtain some indication of the total potential market, it is worth considering each type of photovoltaic system application in turn:

- Small stand-alone systems

    Telecommunications systems, educational television sets, pumps for water supply and irrigation, refrigeration and ice making plants, and remote metering devices are all devices which can be powered by photovoltaics. There will be a growing market for each - the potential market for small

	Estimate 1980	Annual Average 1981–85	Annual Average 1986–90	Average Annual Percentage Growth Rate 1980–90
Electric Power*				
Thermal	8.0	11.8	15.4	9.1
Hydro	9.2	13.5	15.1	6.8
Nuclear	1.2	2.1	8.8	30.4
Other	0.1	0.1	0.4	20.3
Subtotal	18.5	27.5	39.7	10.7
Coal	0.5	0.7	1.5	15.8
Oil				
Exploration	0.5	1.0	1.5	11.6
Development	2.1	2.5	3.2	4.3
Subtotal	2.6	3.5	4.7	8.2
Gas	1.0	1.7	2.7	14.2
Alcohol	0.5	0.9	1.2	12.4
Fuelwood	0.5	0.6	1.3	13.6
Refineries	1.0	1.8	2.3	11.8
Total	24.6	36.7	53.4	10.9

* Includes cost of transmission and distribution.

Table 4.1 Principal investment requirements in commercial energy (1980–90) for oil importing developing countries in billions off US dollars (1980) (Source: World Bank)

solar pumps alone has been estimated as 50 million units with array power from 200 to 1000Wp (67). The market for small photovoltaic refrigerators for use by village dispensaries has been estimated at about 5000 units per year over the next 10 years. (68).

- Larger stand-alone systems

Photovoltaic systems can replace stationary diesel generators and are well-suited for providing electrical power to remote villages. The potential market is very large. Assuming 75% of the estimated 10 million villages are suitable candidates for photovoltaic systems and if the average power required is 50kWp, the total potential market is 3750GWp.

- Grid-connected residential systems

A large potential market for grid-connected residential photovoltaic systems will open up within 10-15 years in developing countries, especially those having the highest solar insolation levels and high cost of grid-supplied electricity. Larger systems serving the service, commercial, institutional and industrial sectors will also become technically and economically viable within a similar time span. The potential market is enormous - the main constraints will be availability of finance and manufacturing capacity.

- Photovoltaic central generating stations

Assuming continued reduction in system costs and real rise in costs of imported fuel, photovoltaic generators of 1MWp and larger will become viable in 15-20 years in many countries in the world's 'solar belt' (ie, between latitudes 30° S and 30° N). Such generating stations could then be constructed by utilities either as large multi-megawatt central stations or as smaller distributed systems, to reduce transmission costs.

Some developing countries will be able to establish their own photovoltaic industry, reducing import requirements to a minimum. For other countries this will not be practicable for many years, but nevertheless a high proportion of the cost of photovoltaic systems can be met by local input. The solar cells and power conditioning equipment may need to be imported, but encapsulation, and manufacture or assembly of the system components and array support structures, are activities that could well be carried out within the country.

In the near term however, in most developing countries all photovoltaic systems will need to be imported and it is therefore important to consider the implications of such an international trade, both from the purchasers' and the sellers' viewpoints.

Factors affecting export activity

Manufacturers of photovoltaic systems who wish to export to developing countries will need to recognise that developing countries have different demands and requirements from the domestic market in Europe or USA. They will need to develop an appropriate market approach that is sympathetic to the needs of developing countries and naturally firms that have experience in doing business with developing countries and have established sales outlets and service networks for other products will be at an advantage.

Many photovoltaic systems will be cost effective in developing countries several years before they are cost effective in the domestic market. For countries in the world's 'solar belt', photovoltaic systems costing about ECU 8 to 12/Wp will be competitive with diesel generators in remote areas, where the true fuel cost today can often exceed ECU 1.00 per litre and the effective life of a small generator may be less than 5000 operating hours, resulting in unit costs well in excess of ECU 0.50/kWh (ie, almost 10 times the cost of grid-supplied electricity in Europe).

As the benefits of photovoltaic systems become established and widely known, demand is likely to grow very fast. Two factors are of great importance when assessing the export potential, namely marketing approach and finance, and it is worth considering these two factors in some detail.

The marketing approach will determine whether an exporter succeeds in a particular market. Developing countries have little experience of photovoltaics and need to be satisfied that these are worth the substantial initial investment. They will not want to be used as a testing ground for new products at their own expense, but with proper collaborative and cost-sharing arrangements they will appreciate the need for field demonstrations of appropriate systems, so that the technology may be evaluated for their own circumstances.

The export market will in general require systems designed specifically for the application and the circumstances that exist in remote rural areas. The exporter must make every effort to determine the real system requirements and the level of technical ability of the eventual operators of the system. This will require the exporter to establish an expertise and data base so that a system can be designed for the application which meets local criteria at acceptable cost. The export of photovoltaic systems cannot be limited to the delivery of crated systems at the port of entry: the supplier will have to ensure that the system is correctly installed, that the owners know how to operate and maintain it and that satisfactory service and follow-up arrangements are made.

International aid agencies and governments are assisting in this respect, by sponsoring demonstration projects and training programmes (see for example projects of this nature being sponsored or assisted by the European Community as listed in Appendix C). Reference has already been made to the UNDP/World Bank project to test and demonstrate solar water pumps for irrigation and drinking water supplies in developing countries, with the ultimate aim of identifying how equipment commercially available today can be improved to become more suitable for use by villagers and farmers in developing countries, with recommendations for local manufacture and assembly (33, 34). The Inter-American Development Bank (BID) is assisting with a solar energy training and development programme in the Dominican Republic (69), and a number of other examples of similar nature could be cited.

A major factor in determining market share will be the extent to which the exporter involves local firms in sales, manufacture, installation and follow-up. The main reasons for this relate to the conditions, both political and economic, in the developing country concerned. Firstly, many governments give preferential treatment to firms which are wholly or partially locally owned and which involve the maximum amount of local input. Contracts will be awarded to such firms even if out-bid by the overseas exporter. Secondly, many developing countries have restricted the range of materials that can be imported in order to conserve valuable foreign exchange and reduce the pressure on the balance of payments. The third reason is that in many countries the administrative procedures to be followed by importers are complex. Local firms will be more familiar with accepted practices and thus better able to achieve results (and obtain payments) without long delays. Finally, there is the important matter of follow-up and the provision of a responsive technical advice and spare-parts service to the users of the photovoltaic systems supplied. A locally-based firm can do this more effectively and their involvement from the start will help increase the confidence of the buyer before he commits himself.

The value of involving indigenous firms in the photovoltaic sales enterprise cannot be over-stated. They can also assist in the important matter of technical information dissemination. It is imperative in a technology as new and changing as photovoltaics that the buyers are kept up to date. Sales will be lost and confidence undermined if decisions are made based on outdated cost and performance data. The use of genuine demonstration projects to help convince potential buyers of the practicality and reliability of photovoltaic systems should be emphasized, since an operational system is a better sales medium than literature packages however well presented.

Finance

Difficulty associated with financing the initial purchase of photovoltaic systems is, and will continue to be, a major factor limiting sales. Many studies have shown that, on a full life cycle basis, and given some further price reduction, photovoltaic systems will be be competitive with alternative power sources, yet it is those countries with the greatest need for such systems that are least able to finance their purchase. Already faced with huge international debts and balance of payments deficits, these countries are finding it difficult to finance existing commitments without taking on additional burdens.

The attitude of the international banking community and the aid institutions will be a crucial factor in the provision of finance for high capital cost, low operating cost renewable energy systems such as photovoltaics. Although finance for the construction of large-scale, conventionally fuelled power stations will continue to be necessary for many years, the European Investment Bank, the World Bank and other funding agencies are expected increasingly to favour projects involving distributed power generators for village electricity supplies and rural irrigation schemes using renewable energy resources such as wind, solar, biomass and micro-hydro.

It should be noted that in many developing countries the domestic prices of petroleum products and electricity are subsidized for social and overall economic reasons. Duties and taxes on imported items such as electrical machines are often high. These factors strongly influence the investment decisions of private individuals and organisations. If properly-based decisions on investments in electricity generation plant are to be made at national policy level, it is important that the true costs of all technologies, free of subsidies and taxes, are used in the economic analyses. The mix of technologies best suited to a country's national objectives, and with a true appreciation of the economic consequences, may then be selected. The implementation of the favoured technologies may then be encouraged by providing low-interest loans, tax credits or leasing schemes.

4.3 Total world market for photovoltaics

Controlling factors

So far in this chapter we have been discussing the potential for photovoltaics in developing countries where undoubtedly the need is perceived to be very great and the potential market immense. Until such time as individual countries establish their own photovoltaic industry, the demand can only to be met by imports from industrialized countries, principally the USA, Europe and Japan.

The potential domestic market for photovoltaics in the USA, Europe and Japan is also high, particularly in the USA, as many studies have indicated. The large sums invested by the European Community, by governments and by industry in photovoltaic research, development and demonstration are evidence of the great hopes being placed in this technology as a major future energy resource. The domestic market in Europe, though less than in the USA, is nevertheless considerable, as discussed in Chapter 3. Japan also sees low-cost photovoltaics as being a major energy resource for both domestic and export markets.

The main factors controlling the rate at which photovoltaic systems will be introduced will in general be the same for all countries. They can be summarized as:

(i)   Technical

Systems must first be appropriate to the needs and be of demonstrated reliability with long lifetime prospects before large sales can be expected. Manufacturers should anticipate rapid market development and invest accordingly.

(ii)   Economic

Normally the photovoltaic system must be shown to have a positive economic net benefit in comparison with the various alternative technologies available to serve a specific application. In some cases, factors such as the need for pollution-free operation or the need to have the security of a source independent of the grid, may restrict the number of competing systems to be evaluated and thus favour the choice of a photovoltaic system.

(iii)   Financial

Even though of demonstrated economic viability, it may be difficult to proceed with the introduction of a photovoltaic system due to the high capital cost and the lack of long-term finance at sufficiently low interest rates. The availability of low-cost finance and the various tax-rebate possibilities open to the potential purchaser are of considerable significance in this respect. Such incentives are well-established in the USA, but are not in general available in European countries.

(iv)   Other constraints

Several other factors, neither technical nor economic, may impede the introduction of any new technology such as photovoltaics, including lack of appropriate regulations, planning obstacles, failure of utilities to co-operate, resistance by Trade Unions active in the existing energy market, and so on. These factors are further discussed in Chapter 6.

Total world market

It is impossible to make precise estimates of the total world market for photovoltaics, there being so many unknowns. Indeed as the technology of photovoltaics changes, other technologies may also be expected to change, thus any estimates made now must be largely speculative. Many studies have been carried out by governments and by the photovoltaic industry to obtain some indication of the possibilities (see for example references 58, 62, 70, 71). One such forecast is shown in Figure 4.2, based on figures published by the US Department of Energy (DOE), which indicates a total installed photovoltaic capacity of some 20 GWp by 1988, rising to between 100 and 1000 GWp by 2000. This forecast was prepared before the current series of budget reductions in the USA and for this and other reasons it is now generally considered that photovoltaic system prices will not fall as rapidly as previously hoped. Consequently, the total capacity forecast is almost certainly far too optimistic.

It should be noted that this DOE forecast assumed an annual doubling of photovoltaic sales every year until 1988, reaching at that time a cumulative total of 4GWp. This rate of increase has certainly been achieved in the last few years, when sales have been strongly influenced by government development policies and purchases supported by funds from international aid agencies. However it is hard to envisage this high rate of growth being sustained in a free commercial market, due not only to the various inhibiting factors already mentioned but also to the long lead times and finance necessary for the construction of the ever-larger factories to produce the photovoltaic systems.

Undoubtedly the factor of dominating importance is module prices. Based on the module price forecast presented in Chapter 2 (see Figure 2.29), a corresponding projection of total world sales has been prepared, as shown in Figure 4.3. Both the module price forecast and the total sales projection curves are based on information obtained in the course of many discussions with photovoltaic manufacturers and specialists during 1981. There is considerable uncertainty regarding the price levels achievable by 1985 and thereafter, and this is reflected in the sales projections. The projection is referred to below as 'CEC 1982 projection of total annual photovoltaic sales worldwide'.

The CEC 1982 projection may be summarised as follows:

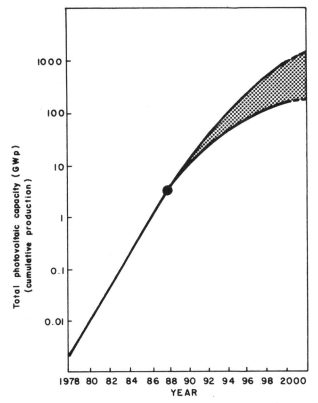

Fig 4.2   Estimate made in USA of total growth of
          photovoltaic systems

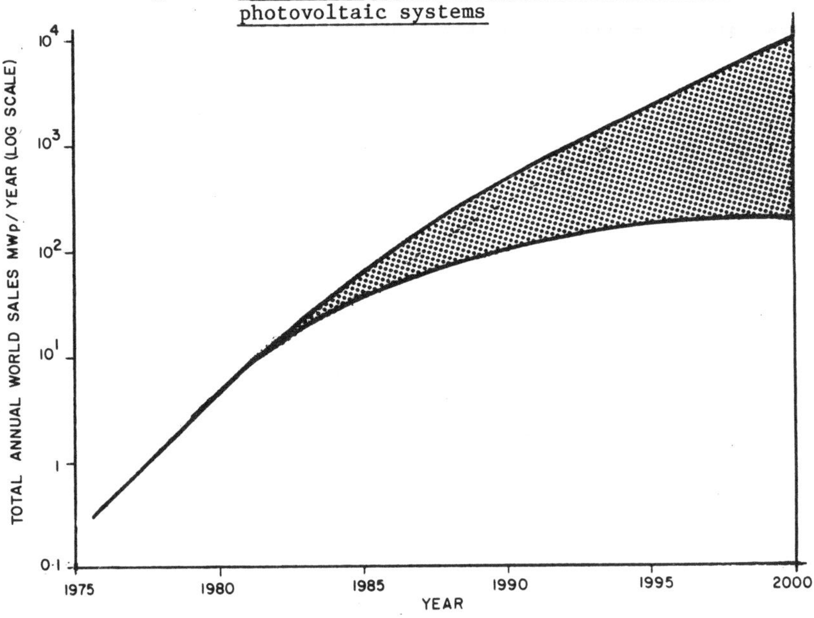

Fig 4.3   CEC 1982 projection of total annual sales of
          photovoltaic systems

Year	World Sales MWp/year	
1980	4	
1985	30 –	60
1990	100 –	400
1995	200 –	3000
2000	200 –	10000

To achieve the higher estimates of sales would require average annual growth rates of about 55% for the period 1980–1985, 45% for 1985–1990 and 40% for 1990–2000. Provided the photovoltaic industry is given adequate support and encouragement (and some suggestions for appropriate measures in Europe are given in Chapter 7), there is every reason to anticipate that these growth rates will be achieved.

To estimate the annual value of photovoltaic systems sales, it is necessary to combine the volume projection, Figure 4.3, with the price projection, Figure 2.29. This results in the following projection for the total value of photovoltaic system sales worldwide per annum:

1980	ECU 120 –	150 million
1985	ECU 250 –	400 million
1990	ECU 600 –	1200 million
1995	ECU 650 –	3600 million
2000	ECU 700 –	20000 million

Even at the lower bound estimates, photovoltaics are thus expected to become very big business. It is now appropriate to consider which companies are involved today and what developments can be expected in future on the supply side of the growing photovoltaic industry.

CHAPTER FIVE - PHOTOVOLTAIC INDUSTRY IN EUROPE

5.1 The international market

The expected growth in worldwide demand for photovoltaic systems will present opportunities for European industry. The main competition is likely to come from two directions: the United States, which already has a well-established photovoltaic industry and an extensive research and development programme; and Japan, where much effort is being directed towards the development of large area, low cost amorphous silicon cells. It is likely however to be several years before the Japanese have a commercially viable product, but when they do no doubt it will be marketted vigorously worldwide. Until such time, the international trade in photovoltaic systems will be largely supplied by Europe and the United States, although some developing countries are already taking steps to establish their own photovoltaic manufacturing capability.

European countries rely on exports to sustain their economies far more than the USA and many European firms have extensive experience and sales contacts throughout the world. There are therefore good reasons for anticipating that the European photovoltaic industry has definite prospects of obtaining a large share of the total world market for photovoltaic systems and the associated engineering services.

Clearly, to be competitive, European industry must be able to market the right product at the right price at the right time. To do this, the technology for components and complete systems must be of the highest quality, striking a balance between advanced concepts and proven designs and materials. The manufacturing techniques and scale of production must be such as will ensure low costs. The product offered to obtain orders must be matched by follow-up and after-sales service. This will require considerable commitment at every level. Every effort should be made in developing countries to involve local industry in the manufacturing and sales process.

To obtain a better appreciation of the quantities of materials involved, let it be assumed that total world sales of photovoltaic systems in 1990 amount to 400MWp (ie, the higher level estimated in Chapter 4, Figure 4.3) and that European industry is able to secure 40% of this market: that is, about 160MWp. System sales would be of the order of 50000 per year. Assuming about one third of the solar cells use advanced Czochralski mono-crystalline silicon technology, one third poly-crystalline silicon and one third ribbon silicon and that modules similar to those made today are used, the approximate quantities of the three main materials needed for the European share of would be:

Solar grade silicon: 3500 tonnes per year

Glass (assuming structural top cover): 20000 tonnes per year

Aluminium for framing: 9500 tonnes per year

Although it is likely that the majority of the solar cells would be made in Europe, it is possible that a significant proportion of the encapsulation would be carried out by associated firms in developing countries. It is worth noting that the amount of solar grade silicon required is equivalent to the current total world production of semiconductor-grade poly-silicon material, of which only about 1% at present is used for photovoltaics.

As the international trade in photovoltaic systems increases, there will be increasing business to be won by companies specialising in components for particular applications such as high efficiency pumps and motors. It may also be expected that there will be a vigorous international trade in electronic power conditioning and control systems, batteries and basic materials, particularly silicon and glass. These are all areas in which European industry can take a prominent share.

5.2 Current status of photovoltaic industry in Europe

Introduction

There is a strong and growing photovoltaic industry in Europe, with some seven or eight major companies and a similar number of smaller ones currently manufacturing and marketing photovoltaic systems and components for a wide range of applications. Several have links with photovoltaic manufacturers in the USA, either as shareholders or as subsidiaries or licensees.

Several European companies have developed products specifically for the photovoltaic industry. One such is Wacker-Chemitronic (FR Germany), a major chemical group which has specialised for many years in the production of mono and, more recently, poly-crystalline silicon wafers. They estimate that they have supplied about 60% of the silicon used worldwide for photovoltaics to date and they are now expanding their semiconductor-grade poly-silicon plant from its present 1200 tonnes/year capacity to about 1800 tonnes/year in anticipation of growing demand. They are also planning to construct a poly-silicon plant at their Portland, Oregon, USA, factory.

The Wacker-Chemitronic subsidiary Heliotronic is conducting research to develop techniques for achieving lower-cost silicon solar cells through such means as:

- solar grade silicon production

- improved crystallization techniques, including research into multi-crystalline sheet growth techniques

- improved ingot slicing and multi-blade cutting of wafers

Wacker-Chemitronic hope to be able to produce poly-crystalline silicon wafers ready to be made into cells for about ECU 1.00/Wp by 1988 compared with the current price of about ECU 2.90/Wp.

Varta (F R Germany) is another company that has foreseen the need for products specifically developed for photovoltaic applications (30). Their Vartabloc battery offers high performance as a stationery battery with low self-discharge and good cycling characteristics. They have also developed a micro-processor based battery charge controller which continuously monitors the battery state of charge and regulates the charge/discharge current accordingly.

Turning now to photovoltaic module manufacturers and system suppliers, the position in Europe is still developing, with new companies being formed and established companies being regrouped. The situation as of end-1981 is reviewed below, country by country, taking the European Community first. Brief reference is also made to the various European Community photovoltaic pilot projects due to be built and operating by June 1983 (72).

Photovoltaic industry in the Federal Republic of Germany

AEG-Telefunken has been active in photovoltaics for some 15 to 20 years and is Germany's largest photovoltaic manufacturer at present, although they are likely to face strong competition from Siemens when their photovoltaic manufacturing plant becomes operational in 1982. AEG has been working for several years with Heliotronic, the Wacker-Chemitronic R&D subsidiary, on the development of low-cost, high-efficiency poly-crystalline silicon cells using Silso material.

A large highly-automated plant for making photovoltaic modules has been built by AEG. At present its capacity is about 20kWp per shift but ultimately an output as high as 1000kWp per shift may be achieved with the same basic machinery. With the introduction and development of this plant, AEG foresee the cost of their photovoltaic modules steadily falling to about DM5/Wp (ECU 2.00/Wp) by 1985, as was shown in Chapter 2, Figure 2.26. AEG expect that further cost reduction after 1985 will be dependent on new technology, such as ribbon silicon cells or thin films (amorphous silicon or another type), as yet still in the research and development stage.

AEG are currently marketing a wide range of photovoltaic systems, including

- hazard beacons
- pumping systems (small and medium sized)
- clock systems
- radio telephone systems for railways and other users
- television sets
- VHF radio and TV transmitters
- telecommunications relay stations
- monitoring stations of various types
- security systems
- power packs of various sizes up to 1000Wp for boats and holiday homes.

AEG have also been involved in providing photovoltaic systems for solar village demonstration projects built or being planned for Argentina, China, Indonesia and Philippines. They foresee pumping systems for drinking water and irrigation in developing countries as being a particularly important market for photovoltaics, along with water purification plants and food preservation systems (refrigeration and ice-making). AEG's total sales of photovoltaics in 1980 are understood to have been about 150kWp and the company reports rapid growth in business, which could possibly rise to as much as 1000kWp in 1982.

Another major company active in photovoltaics is Siemens AG. Although Siemens have been making photovoltaic modules for some time, they have in the past used solar cells made by others. They have now decided to install their own silicon and cell production facilities and to extend greatly their capacity for making modules. The new plant is planned to be operational in 1982 and its capacity is expected to be about 2MWp per year in 1983-1984. Siemens have opted for the well-proven mono-crystalline silicon cell technology and plan to build 120Wp modules, which will be the largest of any manufacturer.

Siemens are developing a number of systems for photovoltaic applications. They are working with KSB, a major German pump manufacturer, on the development of pumping systems from 600 to 2500Wp array output. Three-phase ac motors would be used for submersible pumps in borehole applications, with variable frequency control devices to maximise pumped water output. The first units are planned to be ready for commercial markets in mid-1982.

Siemens are also developing components associated with power conditioning and control systems, such as dc-dc converters and high efficiency dc-ac inverters.

The nuclear engineering company Nukem is now conducting development work in preparation for the construction of a pilot production plant for making cadmium sulphide thin film solar cells. If successful, a much larger plant will be built, producing modules for possibly as low as ECU 1.00/Wp in 1985. Nukem will use the cadmium sulphide thin film technology originally developed at the University of Stuttgart, although certain modifications are understood to be needed for the larger-scale production process involved. Cadmium sulphide solar cells from Nukem are expected to be commercially available by the end of 1982.

One of the European Community photovoltaic pilot plant projects will be in Germany. The 300kWp system for Pellworm Island, when installed will be the largest flat-plate photovoltaic array in the world. AEG-Telefunken are the main contractors for this plant, which will provide power to a tourist centre on the island. The array structure is unusual in that it will be raised up high enough to allow sheep graze underneath. The system will incorporate battery storage and will be operated in conjunction with the grid.

Photovoltaic industry in France

There are many organisations in France engaged in photovoltaic research and development but two main groups have emerged as industry leaders. Radiotechnique Compelec (RTC), the French arm of the Dutch electronics group Philips, has recently joined two other French companies, CGE (Compagnie Générale d'Electricité) and Elf Aquitaine, to form Societé Francaise de Photopiles (SFP), in which CGE has the controlling interest. RTC has handed over its substantial photovoltaic research, development and manufacturing facilities to this new group. Photowatt International, a wholly-owned subsidiary of CGE, is also part of the new group which not only provides financial support but also the benefits of an established world-wide marketing network, particularly through Elf Acquitaine and the RTC/Philips connections.

Photowatt has an associated company in the USA, with its own manufacturing facilities. They are also establishing a company known as Photowatt

Afrique in Abidjan, Ivory Coast, mainly on the strength of a large programme in Ivory coast and neighbouring countries to use photovoltaic generators instead of batteries for educational television sets in the rural areas.

In France Photowatt have introduced a new range of modules incorporating the best features of the previous Photowatt and RTC designs. Cell and module production is being concentrated at the large RTC facilities at Caen and research and development will be concentrated at CGE's Laboratoires Marcoussis. Group management, system design and commercial sales are now being centred on the Photowatt premises in Reuil-Malmaison, near Paris.

In 1980, the Photowatt-RTC combined output of photovoltaics was about 300kWp. The recently-installed lamination equipment at Caen is said to be capable of encapsulating some 5MWp per year of photovoltaic modules.

Photowatt are concerned to ensure that photovoltaic systems are efficient and reliable in order to build up good user experience. They have recently developed a range of pumping systems from 200 to 4000Wp array power which utilise motors and pumps specially adapted for photovoltaic applications. In addition to the normal range of stand-alone applications such as generators for telecommunications, navigation beacons and lighting units, Photowatt are also developing a range of photovoltaic-powered refrigerators for professional users such as medical dispensaries, with capacities 100 to 300 litres.

The other main group in France engaged in the design and manufacture of photovoltaic systems and components is Leroy-Somer, a medium-sized electrical company with some 4000 employees producing ac and dc motors, alternators, heat pumps and control systems. The subsidiary company France-Photon was formed in 1978 and is solely concerned with the production of mono-and poly-crystalline silicon solar cells and modules, using Solarex technology under licence. Sales in 1980 were about 110kWp and they hope to achieve 300kWp in 1982, rising to at least 1MWp by 1985. For production in excess of this level, France-Photon foresee the need for a larger industrial group to cover all technologies starting from silicon material.

The design and marketing of complete photovoltaic systems is undertaken by two other Leroy-Somer subsidiaries: Pompes Guinard, who specialise in solar pumping systems, and Systemes Solaires, who cover the full range of photovoltaic applications. Pompes Guinard have developed a particularly successful range of efficient pumps and motors for use with photovoltaic generators and more than 50 such units of various sizes had been supplied by mid 1981, many with funding from the French government or the European Community.

The French company Total Oil has a controlling interest in a US company called Photon Power, of El Paso, Texas. Photon Power has been operating a pilot production facility making cadmium sulphide thin film solar cells for some time and has recently constructed a plant which will have an initial capacity of at least 5MWp per annum. The cells are expected to be at least 4% efficient, with later improvements anticipated that will bring this up to 6% or more, at costs of less than ECU 1.00/Wp. Provided the Photon Power modules give reliable performance and long lifetime, they may be expected to provide strong competition to established silicon cell technology.

Three of the European Community photovoltaic pilot-plant projects are to be built in France and one on the French Overseas Territory of La Guyane in South America.

The project in La Guyane is being undertaken by Seri Renault Ingénierie (France) in association with Photowatt International, Jeumont Schneider (France) and Varta (F R Germany). The 35kWp system will provide power for a remote village and will incorporate battery storage and a diesel generator for emergency back up.

Photowatt International is the contractor for a 50kWp system to be installed at Mont Bouquet, France, to provide power for radio and television transmitters. The system has the utility grid as back-up but will not be operated in parallel with the grid. Instead, micro-break change-over switching will be incorporated but only two switch operations will be permitted per day as each change-over causes flutter on TV screens. Battery storage will be provided to cover cloudy periods during the day.

Photowatt International is also the contractor for a 50kWp system to be installed on the roof of one of the buildings at Nice airport. The power will serve various airport systems.

A 44kWp stand-alone system will be built by Leroy-Somer in association with France Photon and Oldham Batteries (UK) at Rondolinu Cargese, Corsica. The system will provide power for a village and will incorporate battery storage sufficient for five days supply. A diesel generator will be provided for back up purposes. The load is expected to vary from 50kWh/day in winter to 111kWh/day in summer, when the population is boosted by tourists occupying holiday homes.

Photovoltaic industry in Italy

Solar energy is seen as a major energy resource in Italy and there is substantial government support for research, development and demonstration projects. The CEC-supported Eurelios project, the world's first 1MWp solar thermal power station, is sited at Adrano, Sicily, and there are also plans for a 1MWp photovoltaic plant known as Delphos, to be sited in southern Italy. The government's energy activities are co-ordinated by an interministerial economic planning committee (CIPE). A committee has also been formed between ENEL (the state electricity company), ENI (the state energy company) and ENEA (the state nuclear organisation) to develop and implement new and renewable energy systems. Research and development activities are also supported by CNR (the National Research Council).

Of the many organisations involved in photovoltaic research and development activities in Italy, four are emerging with significant manufacturing capacity either available now or planned.

One of these is Ansaldo, a state-owned electromechanical company appointed by the Government for the development and production of energy systems and power plants (nuclear, thermoelectric, hydro and geothermal), with a complete line of component development. Ansaldo is closely involved with the 1MW Eurelios project and is also building a 35kW hybrid thermal/photovoltaic plant in Australia. Ansaldo has installed a photovoltaic cell and module manufacturing plant based on mono-crystalline silicon technology with capacity of 200kWp per annum. It is intended to enlarge the capacity to 400kWp in 1982 and to 2.5MWp in 1985, at which time a solar grade silicon plant will also be constructed.

Ansaldo has recently acquired a controlling share in another Italian company, Heliosil, an emerging research and development organisation working in the field of solar grade silicon on the basis of proprietary know-how of silicon refining and casting techniques. Ansaldo can thus now cover the complete manufacturing line from material to large and small systems.

A strong group known as Pragma has recently been formed in Italy under the auspices of ENI, the state oil company. Through a share holding in Solarex Corporation and its affiliate Semix in the USA, ENI now has access to Solarex/Semix technology, particularly important for the manufacture of solar grade silicon. ENI has also purchased a controlling share in the Italian photovoltaics company Solaris, a licensee of Solarex. Agip Nucleare, a division of ENI's oil and gas subsidiary Agip, manages most of the projects in the field of renewable energy and will be responsible for this new photovoltaic group. Plans are well advanced for the construction of a 5MWp per year integrated production facility including a solar grade silicon plant, to be ready in 1984. Output capacity in 1982 is about 500kWp per year (on a single shift basis).

Agip designed and built a 3kWp photovoltaic powered reverse osmosis desalination plant near Crotone, in southern Italy, supplying water to a gas dehydration plant. Operation started in 1980. A second similar plant has been built to serve a village on the island of Lampedusa.

There are two other companies active in the field of photovoltaics in Italy: Helios, a licensee of Solec International (USA), and Adriatica Componenti Elettronici (ACE), which is planning to establish production facilities in due course using Siemens technology.

Three of the European Community photovoltaic pilot plants will be in Italy. The largest will be an 80kWp stand-alone system on the volcanic island of Alicudi, to provide power for the 140 residents and the many summer visitors. ENEL working in association with Ansaldo is responsible for this project, which will incorporate photovoltaic modules supplied by both Ansaldo and Solaris. The system will include battery storage sufficient for several days supply. A diesel generator will provide emergency back up. Care has to be taken to reduce the visual impact of the array, which will be built on the terraced slopes of the extinct volcano.

Pragma is responsible for a 45kWp stand-alone system to be built on Giglio Island. The power will be used to operate an ozone generator for sterilising potable water supplies and for a cold store for perishable food stuffs. Battery storage will be provided. Ansaldo modules will be used.

ACE are responsible for a 65kWp stand-alone system to be built on Tremiti Islands to power a reverse osmosis desalination plant. 70% of the photovoltaic modules will be supplied by Siemens and the balance by Ansaldo. Battery storage will be provided of sufficient size to ensure that the desalination plant may be continuously operated throughout the daylight hours of each day.

In addition to the three European Community pilot plants, one other large photovoltaic project is to be built in 1982/83 in Italy. This is a 70kWp stand-alone system to be built by Pragma at Zambelli, near Verona, to pump water from a low reservoir to a higher reservoir. A relatively small amount of battery storage will be provided to cover periods of low irradiance and to stabilize the voltage supplied to the two motors. Grid power will be available as back up.

Photovoltaic industry in the Netherlands

Apart from university based research and development activities, the main organisation actively producing photovoltaic systems in the Netherlands is Holec Solar Energy of Eindhoven, a licensee company of Solarex. Holec have recently decided to construct a 2000kWp per year production facility for cells and modules, based on Solarex and Semix technology. Until such time as that plant is ready, Holec will continue to assemble photovoltaic systems using bought-in components.

Holec are responsible as main contractors for the 50kWp European Community pilot plant on Terschelling Island, Netherlands. The system will operate either in stand-alone mode or in parallel with the grid, supplying power to a nautical training college. Holec propose to incorporate a number of MPPT's (maximum power point trackers), each serving a section of the array field, in order to ensure maximum power is obtained from the array at all times. The system will incorporate battery storage and will operate in conjunction with the grid.

Photovoltaic industry in Belgium

There are three organisations active in Belgium. The first to be established was ENE (Energie Nouvelle et Environnement), a small independent company now making photovoltaic cells from Semix poly-crystalline silicon wafers and encapsulating them into modules. Their capacity at present is about 150kWp per year (single shift basis) and sales in 1981 are expected to be about 15kWp. In addition to selling modules, ENE offer complete photovoltaic systems for such applications as lighting, telecommunications, refrigeration systems and navigation beacons. Power conditioning equipment and other components are bought-in to ENE specifications.

A potentially much larger group has recently been formed, a joint venture between Fabricable and IDE. Although in practice managed like a private firm, IDE (Industrie - Developpement - Energie SA) is 90% owned by the regional walloon authorities. The company was formed in 1974 to work on solar thermal systems, both engineering and hardware production, and added photovoltaics to its activities in 1978. They have built up considerable experience in the design and construction of both thermal and photovoltaic systems, including the largest flat plate thermal collector array, 2100m^2 at Chevetogne, Belgium, and a 1800Wp photovoltaic pumping system installed in Zaire. IDE are currently extending their facilities for making photovoltaic modules and will use solar cells supplied by Fabricable when production starts.

Fabricable is a major cable manufacturer in Belgium and is currently constructing a solar cell manufacturing facility, with technology based on the screen printing techniques developed at the ESAT Laboratory of the Catholic University of Leuven as part of the Belgian national energy research and development programme. The plant was scheduled to be in production by the end of 1981 with a capacity understood to be at least 250kWp per year.

Two of the European Community photovoltaic pilot plant projects are to be built in Belgium. The largest is the 63kWp system for Chevetogne for which IDE in association with ACEC (Ateliers de Constructions Electriques de Charleroi SA) is the main contractor. The system will provide power for the main circulating pumps for a large solar thermal heating installation for a swimming pool, plus lighting and other auxiliaries. Grid power will be available for back-up.

The other pilot plant is an interesting industrial application to be built at Hoboken, Antwerp, where a 30kWp system will provide dc power for a hydrogen production plant and for water pumping. The main contractor is Gensun-C50, a consortium of four large companies:

- Societé General de Belgique (holding company)
- ENI (Electrische Nijverheidsinstallaties)
- Metallurgie Hoboken Overpelt
- Societé de Traction et d'Electricité

Photovoltaic industry in the United Kingdom

There are many organisations involved in photovoltaic research and development in the United Kingdom but at present only two companies are actively engaged in the design, manufacture and marketing of photovoltaic systems. Lucas BP Solar Systems is notable for having won two large contracts in 1980 for the supply of photovoltaic arrays and storage batteries to power telecommunications equipment, one in Colombia worth US $2.75 million and the other in Algeria worth US S1.83 million. Lucas BP market a wide range of stand-alone photovoltaic systems for such applications as pumping units, refrigerators, lighting packages, battery chargers and television sets and offer engineered systems for telecommunications, cathodic protection and other special applications. Their policy is to supply complete systems rather than photovoltaic modules alone. At present they purchase solar cells made by other companies and encapsulate them in modules of their own design. Later, when sales volume increases, Lucas BP have said that they would consider establishing their own cell production line. Meanwhile, planning studies have been made for the installation of a large-capacity automated module production line, incorporating robot systems.

Lucas BP is owned jointly by Lucas Industries and British Petroleum and this gives them not only financial support but also the advantage of a well-established world-wide marketing network through the parent companies. Lucas is a major electrical component company specialising in equipment for the international automotive industry. BP is a major oil company and is already extensively involved in many countries in the manufacture and marketing of solar thermal systems for water and space heating.

The other company actively marketing photovoltaic systems in the United Kingdom is Solapak, a company associated with Solarex Corporation (USA). Solapak design systems and build, through an associated company, their own power conditioning and control equipment. Photovoltaic modules are imported from the USA and made up into systems for various applications. Solapak has recently been reorganised with strengthened management and technical resources and has secured a number of important orders in the Middle East.

In addition to these two companies, it should also be noted that the British glass-making group Pilkington Brothers now own an 80% share in Solec International, a medium sized photovoltaics manufacturer in the USA. This association gives Solec an international partner and Pilkingtons a foothold in the photovoltaics industry to supplement its substantial interests in solar thermal systems. Solec currently has a production capacity of about 200kWp per year. Pilkington Solar Products Limited are already marketing a range of solar thermal systems and they intend to introduce marketing outlets for photovoltaic systems throughout their overseas sales network including most European countries.

One of the European Community photovoltaic pilot plants will be built in England, at Marchwood near Southampton. Lucas BP are main contractors for the 30kWp plant, which will be used to investigate operation of a photovoltaic system under stand-alone and grid-connected modes with various types of load.

Photovoltaic industry in Ireland

So far no manufacturer has established photovoltaic manufacturing facilities in Ireland, although a number of attractive financial incentives are available to encourage the establishment of new industry of this nature. The Microelectronics Research Centre of the University of Cork has however been active in photovoltaic research for many years and this has led to the appointment of the University of Cork in association with AEG-Telefunken (F R Germany) and the Electricity Supply Board (Ireland) as contractors for the European Community photovoltaic pilot plant to be built on Fota Island. The 50kWp system will serve a dairy farm, which is an application where electrical demand is well matched to insolation. The system will include battery storage and will be operated in conjunction with the grid.

Photovoltaic industry in Denmark

As yet, no photovoltaic industry has been established in Denmark, but detailed design work was carried out by DET (Dansk Energi Teknik) in association with Siemens (F R Germany) and Varta (F R Germany) for one of the European Community photovoltaic pilot plants which was to be built for the village of Vester Beogebjerg. Unfortunately this project has had to be cancelled.

Photovoltaic industry in Greece

Although there are no established photovoltaic manufacturing facilities in Greece, there is considerable potential for photovoltaic systems because of the favourable climate and the large number of relatively isolated communities on islands or in the mountainous regions, where the true cost of diesel generated power is currently often as high as ECU 0.50/kWh. The Public Power Corporation of Greece foresees photovoltaics as potentially being an important energy resource particularly if combined with wind generators. There are also some 700 remote lighthouses in Greece, of which many are likely to be suitable for photovoltaic stand-alone systems.

Two of the European Community photovoltaic pilot plant projects will be built to serve Greek islands. There will be a 100kWp system built by a consortium headed by Siemens (F R Germany) on Kythnos. The other members of the consortium are the Public Power Corporation (Greece) and Varta AG (F R Germany). The system will operate in parallel with the existing diesel generators and several small wind generators.

Seri Renault Ingenierie (France), also working in association with the Public Power Corporation and Varta AG, will be constructing a 50kWp stand-alone system to provide power to Aghia Roumeli, Crete, which at present has no electricity supply. This small village receives large numbers of day visitors during the tourist season and the availability of electrical power will significantly improve the income potential for the villagers. A diesel generator will be incorporated in the system for emergency back-up.

Photovoltaic industry in Spain

Spain is one of the European countries with particularly good potential for photovoltaic systems on account of both the solar climate and the number of potential stand-alone applications for remote houses. Government and industrial bodies are sponsoring development work. Presently Copresa, a company in the Philips-RTC grouping, is manufacturing PV modules and mono-crystalline silicon solar cells are being made by Piher-Semiconductores of Barcelona.

A 100kWp photovoltaic generator project is currently being planned by Centro de Desarrollo Tecnologico e Industrial (CDTI) and Standard Electrica, with funding from CDTI, the Ministry for Industry and the electrical utilities association UNESA. The project will use Piher solar cells. Batteries and other auxiliary equipment will be supplied by Tudor and Femsa. A site for this flat plate generator has yet to be selected.

Photovoltaic industry in Switzerland

Plans have been announced for the first fully integrated photovoltaic production plant in Europe to be built by a company known as Pasan at Nyon, Switzerland, with financial financial backing from the Swiss industrial group Hasler AG, two state controlled utility companies and private investors. A licensing agreement with Solarex (USA) provides Pasan, through its Photonetics subsidiary, with advanced photovoltaics production technology, including the Semix technology for making poly-crystalline silicon cast ingots from solar-grade silicon. The ingots will then be sliced into wafers, which in turn will be made into cells and encapsulated into modules. System design capability is also included in the group. Planned capacity is understood to be at least 2000kWp per year for the initial plant.

This development could prove to be of considerable significance for the European photovoltaic industry, since a fully integrated factory for photovoltaics using the latest techniques offers the possibility of achieving low-cost modules and complete systems competitive with manufacturers in the USA and other European countries. The implications for export are also important, as Switzerland has great experience both in exporting electrical equipment worldwide and in the provision of long-term finance with associated government guarantees.

Photovoltaic industry elsewhere in Europe

As far as it has been possible to determine, there are no companies making photovoltaic components in other Western European countries. There is no evidence of Eastern European countries producing components or systems for export or for their domestic terrestrial market.

It can however be expected that once a sizeable market is evident, for domestic and export applications, most European countries will establish their own photovoltaic industry.

Retail Trades and Distribution

Almost all of the photovoltaic systems installed and operating are either demonstration plants or have been installed by large organisations such as oil companies and communications operators who are well equipped to specify their needs and buy direct from the manufacturers. It is therefore still too soon to expect to find the type of product distribution network which is associated with more mature products. If photovoltaic products are to be sold in the quantities suggested in this and other reports, particularly stand-alone and residential systems, the development of effective distribution systems will be an essential pre-requisite.

It has been noted that many photovoltaic manufacturers are supported by large national or multi-national organisations. These will have the ability to draw upon their existing distribution systems in many countries, whether they be local subsidiaries or agents. The more independent photovoltaic manufacturers will need to establish such networks very quickly in order to be competitive as large volume sales begin.

Channels for the distribution of photovoltaic products are likely to include:

- sales to manufacturers of other products for inclusion in complete systems distributed through their existing networks

- sales direct to utilities or major industrial users who may produce their own specifications

- sales direct to users through the manufacturer's agents or own distribution network

- sales through retailers (perhaps via wholesalers) such as existing electrical, plumbing or home-improvement specialists or even new specialist stores offering systems from different manufacturers.

- sales to house builders

- in developing countries, sales through governmental development agencies, to farmers or villages with financial support.

In many countries there are now agents or distributors appointed by manufacturers but so far the proportion of the total sales of photovoltaic modules made as a result of the existence of these arrangements is relatively small.

Clearly as the industry develops there will be a great need to develop the distribution systems. This development will include the education and training of the staff in contact with potential purchasers as an essential sales-aid. Adequate training for the technicians and skilled workers needed to install and maintain photovoltaic systems may also be regarded as part of the development of the retail trade.

Engineering Services

So far the knowledge and expertise gained in the design and implementation of photovoltaic systems has been concentrated in manufacturers and the research and development agencies, many of which are government operated or funded. As the application of this new technology becomes more widespread there will be an increasing demand for advice and assistance regarding the engineering and the economics of systems. Some manufacturers may continue to engineer tailor-made systems for their clients but others will wish to concentrate on the manufacturing and leave the engineering services to others. A demand for suitably qualified and experienced consulting engineers, operating independently of the manufacturers, can be foreseen.

It is particularly relevant that in developing countries independent advisers are often sought to advise and train local personnel. It has been the experience of the construction industry that the import of knowledge from a particular region produces a disposition towards the products of the same region. Thus the creation of a source of advice and information can be the forerunner of a large share of the market for products.

5.3 Photovoltaic industry in the USA and Japan

To see the European photovoltaic industry in a world-wide context, it is important to review the situation in relation to the two countries which are investing the largest sums in photovoltaic research and development, namely, the USA and Japan.

There are currently some 10 to 15 companies in the USA actively manufacturing photovoltaic modules and components for systems. The principal ones are:

- Solarex and Semix
- Solar Power Corporation
- Arco Solar
- Photowatt International
- Applied Solar Energy Corporation
- Solec International
- Solenergy
- Solavolt
- Tideland Signal

These companies and several others are involved in the design and marketing of systems. At present most use the well-established mono-crystalline silicon solar cells except Solarex who are beginning to move over to Semix poly-crystalline silicon cells. Solarex are one of the market leaders and are unusual in being among the few that are independent of external control: for example Exxon own Solar Power Corporation, Atlantic Richfield own ARCO, CGE of France own Photowatt, Pilkington Brothers of the United Kingdom own 80% of Solec and Motorola and Shell Oil are behind Solavolt.

In addition to the companies at present marketing photovoltaic components and systems, there are many more who have been developing new materials and processes that are close to becoming commercial. In this category may be mentioned the following:

Crystalline silicon technology:

- Crystal Systems (HEM process)

Ribbon silicon technology:

- Mobil Tyco Solar (ribbon)
- Westinghouse (web)
- Honeywell (silicon-on-ceramic)

Cadmium sulphide thin film technology:

- Solar Energy Systems - SES (majority owned by Shell Oil, USA)
- Photon Power (owned by Total Oil, France and Libby-Owens-Ford, USA)

Amorphous silicon technology:

- RCA
- Energy Conversion Devices (ECD)

In addition to the companies mentioned above, there are many more engaged on research and development work covering many special aspects. The Jet Propulsion Laboratory manages the low-cost solar array project. Sandia Laboratories co-ordinates work on concentrator photovoltaic systems. SERI (Solar Energy Research Institute) conducts basic research and manages various aspects of the photovoltaic development programme.

Photovoltaic research and development in the USA has until recently been strongly supported by the US Department of Energy. The DOE Photovoltaics Program had a budget of US $119 million in 1979 and $150 million in 1980.

The 1981 budget has been cut back somewhat to $139 million from the $160 million originally proposed and the 1982 budget may be reduced to $63 million from the $161 million proposed, funds being concentrated on longer-term R&D. Despite these budget cut-backs, the US photovoltaic industry is now very strong and is investing from its own resources at a comparable or even higher level than the Federal Government. Although there may well be some casualties due to the reduced budget, it is generally considered that the industry can now stand on its own and thrive, provided fiscal and other obstacles are not introduced that discourage firms and individuals from purchasing photovoltaic systems.

In Japan, as in the USA, there has been a broad-based research and development effort to find new and renewable sources of energy to reduce the nation's dependence on imported oil. A vigorous photovoltaic research effort has been mounted as part of the Sunshine Project and the following targets have been set (73):

- Annual production of photovoltaics: 50MWp/year by 1985
                                      3GWp/year by 1990
                                      20GWp/year by 2000
- Module price of ¥ 50-100/Wp (ECU 0.20 - 0.50/Wp) by 1990

These figures require an annual rate of growth of 100% from 1980 to 1990 and a 20% annual growth rate thereafter. The programme targets are thus comparable with the US DOE targets.

The New Energy Development Organisation (NEDO), a part of the Ministry for International Trade and Industry (MITI), is funding six companies to develop solar cell technology. The companies are: Sanyo Electric, Fuji Electric, Mitsubishi Electric, Teijin, Kyoto Ceramic and Komatsu Electronic Metals. For the first time, NEDO is also directly funding work at universities, including the universities of Tokyo, Kyoto, Osaka, Kanazawa, and Hiroshima. A major programme of supporting development and system demonstrations is also being planned by NEDO. Construction of photovoltaic systems for single-family homes (3kWp), multi-family homes (20kWp), schools (200kWp), factories (100kWp) and central generators (200 and 1000kWp) was started in 1981. A photovoltaic/thermal hybrid system is also under construction, with peak output 5kW (electrical) and 25kW (heat).

With the low-cost targets clearly in mind, the Japanese photovoltaic research and development programme has been concentrated particularly on low-cost crystalline silicon and amorphous silicon techniques. There have been some important technical achievements reported and steady progress is being made towards the targets. Significant features of the Japanese photovoltaic industry include the following:

- Sanyo, Mitsubishi and Fuji are all working on the development of larger area amorphous silicon cells, deposited on glass or stainless steel substrates. For 100m x 100mm cells, it is reported (16) that Sanyo have achieved efficiencies of 5.6%, Mitsubishi 5.03% and Fuji 4.7%. For smaller 10 x 10mm cells, Fuji have achieved 7.8% and Osaka University 7.72%.

- The Teijin company is pursuing another low cost approach, which is to deposit amorphous silicon films on a flexible polyamide film. An efficiency of 3.6% has so far been achieved by this interesting approach.

- The Kyoto Ceramics group is developing amorphous silicon cells deposited on ceramic tiles or bricks, which open up interesting possibilities for the construction industry. Kyoto Ceramics are already marketing modules made up of polycrystalline silicon cells.

- The market for small area amorphous silicon cells has grown very rapidly recently. In the Summer of 1981, Hamakawa reports that about 420000 pocket calculators per month (40% of total Japanese output of calculators) were being produced with solar cell power sets (16). This market is expected to grow worldwide, not only for calculators but also for watches and portable radio/cassette players. Sanyo are market leaders with their 'Amorton' cells giving typical efficiencies over 5%, with particularly good response to artificial light from fluorescent tubes.

- Photoelectron Industries have a production capacity of about 140kWp per year which is now being extended. The present range is based on mono-crystalline silicon technology, with efficient production techniques used in the manufacturing process.

- Shikoku Electric Power Company has installed systems with a total capacity of 25kWp using mono-crystalline silicon cells and plans to increase installed capacity to 1000kWp by 1986.

- Toshiba Electric are understood to be planning a large plant using the ribbon silicon technology they have developed.

- Both Hitachi and Sharp already have production facilities and they are understood to be now planning to build large plants using highly automated equipment to reduce costs.

It seems clear that Japanese photovoltaic manufacturers will soon be in a position to present a serious challenge to US and European manufacturers in overseas markets.

5.4 Value of photovoltaic business in Europe

About 4MWp of photovoltaic systems were supplied worldwide in 1980, worth about ECU 150 million. Although firm figures for 1981 are not available, it is believed that total sales were over 6MWp, worth about ECU 200 million. European manufacturers secured about 12% of the market, the balance being largely supplied by US manufacturers, with small amounts from elsewhere.

European manufacturers expect to produce over 2MWp in 1982, which would be about 15% of the total world market. As was discussed in Chapter 4, the world market is expected to grow to 100-400MWp is 1990 and 200-10000MWp in 2000. What proportion will be secured by European manufacturers is an open question. The potential is there, but major support and encouragement, from both public and private sources, will be needed to ensure a healthy and successful European photovoltaic industry.

European manufacturers should certainly secure the majority of the European market plus a significant share of the export market. A reasonable target would be at least 30% of the world market (ie, double their present share), in which case the European share of the market, based on the upper bound projections given at the end of Chapter 4, would be as follows:

1981	0.8MWp	worth	ECU	35 million
1985	30MWp	worth	ECU	120 million
1990	120MWp	worth	ECU	350 million
2000	3000MWp	worth	ECU	6000 million

Further growth will continue until well into the next century, establishing photovoltaics as a major industry in its own right.

It is of interest to compare these estimates of the value of photovoltaic business with those for low temperature solar thermal systems. In 1981, about 219000m^2 of thermal collectors were supplied within the European Community, the total value of the associated systems being about ECU 110 million. In 1982, it is already clear that this level of business will be less, whereas photovoltaic sales are firmly expected to reach 2.0MWp, worth some ECU 80 million, a level of business already comparable with that for solar thermal systems.

CHAPTER SIX - PHOTOVOLTAICS IN SOCIETY

6.1 Social and industrial implications

Introduction

The biggest barrier to the wide-spread introduction of photovoltaics as a source of power is undoubtedly economic. Although photovoltaic systems have advantages in that they require no fuel, need little maintenance and present no pollution problems, it is clear that until systems are available that, on a full life-cycle costing basis, show net positive benefits compared with alternative energy sources, the market will be restricted to a limited range of specialist applications and government-sponsored experiments. All the indications are, however, that through technological advance and mass production, the break-even costs for photovoltaics to compete first with small diesel generators supplying individual users and then with larger diesel units serving village communities will be achieved within 10 years. There is good reason to anticipate that photovoltaic systems eventually will be available at costs low enough to compete with centrally generated electricity distributed via the grid.

Assuming that technical and economic development does proceed as anticipated and that in consequence the very large markets discussed in chapter 4 do open up for photovoltaics, what will be the implications for society and, in particular, for Europe? For example, what will be the effect on employment and industry, does its introduction require new laws and controlling regulations, what response from the electricity and other energy utilities will be needed and what are the implications for the environment and what are the implications for the individual consumer?

Material and resource constraints

A large-scale photovoltaic industry will require large resources for the manufacturing and installation processes. These resources may be separated into materials and energy. At present there is considerable uncertainty, as the nature of the main technology that will form the basis of the photovoltaic industry in the future is not yet known. Nevertheless, by making certain assumptions, it is possible to derive useful indications.

The diffuse nature of solar energy makes necessary the provision of large photovoltaic arrays to supply significant amounts of electricity. The materials required can be separated into those special to the photovoltaic cells, those needed for the modules that make up the array and those needed for the array support structures.

Considering first the solar cells, material requirements will depend on cell type, efficiency, thickness and whether some form of concentration is incorporated. The material used for commercially available photovoltaic systems today is silicon in very pure (semi-conductor grade) crystalline form in wafers about 250μm thick, but developments are now under way to introduce cheaper solar-grade silicon made in plants built specifically for this purpose. At present, about 15 tonnes of silicon are needed to make 1MWp of photovoltaic cells but this will change in future as more efficient wafer cutting techniques or ribbon processes are introduced. A more important development may be the introduction of thin film amorphous silicon or cadmium sulphide cells, less than 5μm thick for amorphous silicon and only 5 to 8μm thick for cadmium sulphide.

Silicon is one of the most abundant elements on earth and presents no problems of availability although current production rates of purified silicon will naturally have to be increased to meet demand. If cadmium sulphide photovoltaic cells become the dominant technology, then it has been calculated that 5000-22000 tonnes per year of cadmium will be needed for an annual production of 20GWp, on the assumption that the film thickness ranges from 5 to 8μm and the efficiency ranges from 5 to 9 percent. The CEC 1982 projection presented in Chapter 4 as Figure 4.5 indicates that world production could reach 20GWp per year in the late 1990s. Production is expected to continue to rise thereafter, to make photovoltaics one of the world's major industries.

Proven world reserves of cadmium are 700000 tonnes. According to a report made in 1978 by the US Office of Technology Assessment, domestic supplies of cadmium will be sufficient to supply annual production rates in excess of several thousand megawatts a year.

A mixture of cell types will reduce the demand for any single element and thus reduce any risk that material shortages will limit photovoltaic production. Probably greater impact will come from the materials used to encapsulate the solar cells into modules and to support the arrays. It is estimated that about 0.6 million tonnes of tempered glass, 1.6 million tonnes of aluminimum and 250000 tonnes of oil (to make the vinyl pottant) would be needed to encapsulate 20GWp of photovoltaic silicon cells. Estimates of the materials needed to support the arrays are more problematic, since wherever practicable roof structures will be designed to support the arrays directly. Elsewhere, suitably treated timber may provide the most economic solution.

Glass supplies may present some temporary restraint not because of raw material availability but because of the long lead time needed to expand production of tempered sheet glass to match potential growth in photovoltaic sales. The requirements for aluminimum will constitute a relatively small proportion of world production.

Energy

The production of photovoltaic arrays will require the input of energy which, at least during the early development of the industry, will need to come primarily from conventional non renewable sources. An important consideration is the energy pay-back time; that is the time needed for the photovoltaic system to generate the amount of energy that was required in the manufacturing process from raw materials to finished product.

There have been several studies of this topic and estimates of the photovoltaic energy pay-back periods for existing technology range from two to ten years depending on solar insolation, cell efficiency, manufacturing process and array support structures (74, 75). For comparison, some nuclear power stations have an energy pay-back period of about seven months and coal and oil fired plants rather less (considering only the energy used for their construction).

Silicon solar-cell production techniques at present require large amounts of energy, particularly in the refining and purifying of silicon. The energy pay-back period is currently several years but new techniques are being introduced by some manufacturers that are expected to reduce the payback period to less than one year. The concept of the 'solar breeder' factory, which uses the energy from solar cells to run the factory and thereby make more solar cells, then becomes viable. A factory of this type with a 200kWp photovoltaic generator is now being constructed by Solarex in the USA.

The advanced thin film solar cell techniques hold the potential of reducing the energy pay-back period to a few months. Provided the energy pay-back period is less than one year, then it is theoretically possible to increase production of photovoltaic systems at a very high rate without significantly affecting the energy supply situation elsewhere, although some regional imbalances would follow.

The normal energy pay-back analysis does not take account of the different availability and economic values of energy, be it thermal versus electrical or energy to meet peak demand versus energy for base loads. The distinction is important because photovoltaic systems can be used to supply power during peak demands though they have been manufactured with off-peak power. Furthermore, photovoltaic systems may be manufactured in places where energy is relatively cheap and plentiful (eg, using hydroelectric power) and then transported for use in regions where energy in conventional forms is scarce and expensive.

Energy pay-back periods need to be borne in mind when considering the potential for photovoltaic systems as fossil fuel savers. A photovoltaic system may be expected over a period of 20 years to save the equivalent of about 6 to 8 litres of oil for every peak watt installed. The net saving however must however take into account the difference in the energy pay-back periods between the photovoltaic system and the alternative generating systems with which it is being compared.

Employment

It can be expected that any major technological change in the energy sector will affect the level and nature of employment in that sector. For photovoltaics, it appears that the manufacture, distribution and servicing of these systems will affect the number and types of jobs required.

At the manufacturing end, new continuous cell fabrication techniques and automated assembly will greatly reduce the labour intensive methods currently employed in the photovoltaic industry. According to Neff (76), the total labour intensity of photovoltaic systems may be in the range of 5 to 25 million person-hours per GW-year. By comparison, Neff's estimate for direct labour requirements for electricity from coal is between 2.5 and 3.6 million person-hours per GW-year. Although there may be some duplication in these figures, it seems clear that photovoltaic systems will require at least as much and probably more labour than conventional power generation. The nature and type of the jobs involved will however be very different. Instead of the preponderence of manual trades in coal mining and civil engineering construction, there will be greater emphasis on light industry and building trades, with photovoltaic factories well distributed throughout the community and many relatively small companies engaged in design and installation work.

This is not to imply that increased activity in the photovoltaic sector will result in jobs lost in the conventional industries, at least not in the next 20 to 50 years, since most energy scenarios anticipate growth of the coal and other energy industries over that period. Photovoltaics along with other renewables will be deployed mainly to meet rising energy demand and to meet that part of the demand not previously supplied with power from conventional sources.

6.2 Legal and regulatory factors

Legal framework

However interesting or attractive a specific technology, it will be governments and in particular the laws and regulations that they introduce that will provide the context for and determine the rate at which that technology is developed and introduced into society. Miller begins his extensive review of legal obstacles to decentralised solar energy technologies with the following paragraph (77):

New products, even when technically practical and economically competitive, must still overcome many obstacles before becoming successful. This is true for solar energy technologies as well as for more mundane consumer goods; indeed, given the complexities of energy delivery systems, the non-technical obstacles to decentralised solar energy technologies may be especially great. Consumers must be satisfied that systems are reliable and durable; the rates to be charged for auxiliary energy must be determined; building code practices must be settled; and a myriad of other potential problems must be overcome, some of which may not yet be recognised.

Many of these issues will arise as legal provisions, covering the manufacture, sales and performance of photovoltaic systems. Regulations will determine how and where systems are installed, financial arrangements, fiscal concessions and public utility requirements, to mention just some of the aspects affected. Most of these issues are not unique to renewable energy systems, but in one respect solar energy does require special treatment, because of the vital need to have uninterrupted access to sunlight, the 'right to light.'

Solar access

The basic requirement for access to sunlight is obvious but the extent of the legal problem is not always clear. All solar energy systems, whether photovoltaic or thermal, are sensitive to shading from trees or adjacent buildings. A potential purchaser needs to be confident that his neighbours are not going to build structures or allow trees to grow that will reduce the effectiveness of his expensive solar energy system. A few such instances, well publicized, could have a profoundly depressing effect on the market.

The legal framework in many countries includes the concept of nuisance and it could be held that obstructing the passage of sunlight to a neighbour's solar array might be considered a nuisance and, if necessary, be restrained by legal action. This might serve to prevent frivolous or deliberately provocative activities, but it is hard to expect this principle to be effective in preventing landowners from proceeding with otherwise serious and sensible developments on their land, even if these do overshadow a neighbour's solar array at certain times of the year.

Some countries do have laws that restrict new development in front of windows. In the United Kingdom for example, there must be an unobstructed space at least 3.6m outside the window of a habitable room, open vertically to the sky. This principle would not be adequate, even if it were held to be applicable, for solar arrays.

Possible solutions to the access to sunlight problem have been proposed. These include:

1. Zoning: all new development within a defined area must preserve the solar access rights of neighbours, forming solar energy districts.

2. Individual lot protection: this would clarify and strengthen the concept of nuisance, preventing neighbours from causing any obstruction to an existing or planned solar array, by means of a specific restrictive covenant applied to new development or negotiated with existing neighbours.

3. Legal right to solar access: this approach would make access to solar energy a general property right, subject to prior appropriation by existing construction.

Method 3 would be the widest in its effect but difficult to administer in practice. It may be possible to define more clearly its application by including the requirement that existing or potential solar arrays could be registered with an authority which would then protect solar access to them in the event of future building developments. There is a clear analogy with water law (riparian rights), where landowners may register their use of existing water resources on the land, and neighbours are then not permitted to abstract water to the detriment of the registered user.

Development restrictions

The largest impediment to solar installations may well be development restrictions applied either by the local municipal authorities or by privately created architectural controls. Private architectural controls are frequently used to maintain the harmonious appearance of a community intended by the original developer. Residents in such communities may not make changes to the external appearance of their houses without the consent of a designated authority. If the introduction of solar energy systems is perceived to be in the national interest, it may be necessary for governments to legislate that in future no private architectural controls shall be permitted that prohibit the use of solar energy by residents in a community.

A more general issue is raised by the need in many countries for all new development, including external modifications to existing buildings, to be approved by the local planning authorities, who may have implicit if not explicit policies that militate against solar energy installations. It should be recognised that a solar photovoltaic or thermal array can be unattractive although much can be done to reduce the visual impact with sensitive design and detailing. Indeed it is possible to design the arrays such that they become an integrated and positive feature of the building or landscape. The orientation of buildings and the need to preserve solar access rights are two factors that must be appreciated by planning authorities if the introduction of solar energy systems is not to be hindered.

Some examples of areas in which photovoltaic installations may conflict with existing planning policies and regulations have been identified as follows (78):

Height Restriction - roof mounted array projects above allowable height.

Side Yard and Set Back Restrictions - array on ground placed too close to lot boundaries.

Density of Lot Area Coverage Restrictions - area of array on ground plus buildings exceeds the legal building-to-lot area ratio.

Accessory Use Limitation - subsidiary structures may be prohibited or placement limited.

Use Regulations - based on nonconforming use, additions to structures may not be permitted; residentially zoned lot may not be allowed to produce electric power.

Aesthetic or Architectural Controls - design, style materials or colour may be perceived as unallowable when such restrictions exist.

Guidance from governments, possibly even legislation, will be needed to establish policies for dealing with possible conflicts between developers wishing to build solar energy systems and planning authorities.

Standards and codes

The development and diffusion of photovoltaic systems will be affected considerably by the applicable building codes and industry standards. Such codes are designed to protect the health, safety and welfare of the public. Code changes usually require a lengthy and time-consuming consultative process and often no single product approval procedure or agency exists.

Standards and building codes are necessary to protect the public and the industry from poor products and exaggerated promises, but they can add to the cost of products and stifle innovation. Although often viewed as a 'necessary evil', codes can also help the industry. For example, some districts in the USA have requirements that provide for all new single-family homes to include provision for the future installation of a solar energy system.

Photovoltaic systems will have to comply with all relevant building codes, especially those relating to roof construction, fire resistance and electrical safety. The photovoltaic industry in Europe and the USA is developing standards specifically applicable to photovoltaic cells and modules and the following two current documents have already been issued by the Ispra Joint Research Centre of the Commission of the European Communities:

- Standard procedures for terrestrial photovoltaic performance measurements, Specification 101, issued 1980.

- Recommendations for qualification testing of terrestrial photovoltaic modules, Specification 501 issued 1981.

On the matter of safety standards, the Jet Propulsion Laboratory in the USA, as part of the Low-Cost Solar Arrray Project, has recently issued 'Interim Standard for Safety: Flat-Plate Photovoltaic Modules and Panels; Volume 1, Construction Requirements'. Volume 2, Performance Requirements, is in preparation.

A number of areas in which photovoltaic systems may be in conflict with existing codes, have been identified (78). Examples include the following:

Special hazards - such as battery storage

Roof structures - there may be height or area restrictions as well as loading conditions to be satisfied; glass structures may be classed as skylights which require wired glass.

Wiring methods - rigid conduit may be required.

Location of generators - interpretation of whether a PV array is a 'generator' for the purposes of regulating its location.

Until appropriate standards for photovoltaic installations are generally available, care must be taken when applying existing standards not written with photovoltaics in mind. Misdirected standards may cause certain product attributes to be emphasized whilst neglecting others. It is therefore imperative that prospective photovoltaic manufacturers and system designers work closely with the representatives of users and code-drafting bodies to generate and apply relevant industry standards.

Warranties and consumer protection laws

A warranty is a form of insurance which not only raises consumer confidence in the product but gives him some redress in the event of failure. A basic warranty is normally found in national civil laws relating to the quality of merchandise or in contract law, and this is often supplemented by consumer protection law, or by express warranties offered by or demanded of suppliers. The cost of the warranty is spread among all users and can be considerable for a new product, especially if, under consumer protection law, the warranty is required to be certified by an independent agency. Warranties are usually required to cover the materials and workmanship for a specified period of time. Conflicts can sometimes arise if the warranty period distinguishes between manufacturing defects and installation defects. This can become a difficult problem if, as is often the case, the installer is not the same as the manufacturer and each blames the other for the failure.

A more basic issue when considering a new industry such as photovoltaics is whether the imposition of onerous warranty terms (to protect consumers from failure of new products) will put unacceptably heavy burdens on a fledgling industry. Small businesses are often the most innovative yet they lack the resources to pay for high cost warranties. Some risk-sharing between suppliers and customers is indicated — the difficulty is to decide where to draw the line. Another basic question is whether warranties that provide essentially for faulty materials and workmanship are really an adequate substitute for performance guarantees. From the customer's viewpoint, there is no advantage from the manufacturer replacing faulty parts if the system does not work properly anyway. What is needed is a performance guarantee, which might include provision for some measure of reduced output with time due to degradation, plus a warranty covering defective parts and workmanship. This approach is being followed for the European Community photovoltaic pilot plants.

During the period of evolution of standards, consumer protection can probably best be achieved by appropriate education. Governments and the industry can help potential customers evaluate for themselves the implications of installing a photovoltaic system and give guidance regarding appropriate features for the application in mind. A growing list of satisfied customers will do more for the emerging industry than pages of fine print limiting the contractor's liabilities.

A possibility for further support is for governments to help in funding a warranty scheme or performance insurance in support of approved systems. If the wider social and political benefits of solar energy are recognised and accepted by governments, this could be a cost-effective method of encouraging consumer confidence in the new industry without over-burdening the manufacturers with additional overhead expenses. Clearly codes and standards would need to be established to provide a recognised basis for systems to be approved.

6.3 Implications for electricity utilities

Utility/photovoltaic interface

After several years of exhortations from governments, electricity authorities, gas utilities and oil companies to the effect that we should all save energy, a paradoxical situation has now arisen. We are indeed beginning to save energy by becoming more efficient users but, far from applauding our small success, the utilities complain that sales have fallen and that there are now ample stocks of coal and a glut of oil. The result of course is that unit prices must rise yet faster to reduce losses.

No doubt the current economic recession is largely to blame for falling (or at least, not rising) energy demand in industrial countries. Efforts to introduce more energy efficient designs and techniques must continue, but even so it is probable that total energy demand in industrial countries and certainly in developing countries will rise in future. Nevertheless, even with rising energy demand, it is unrealistic to expect any but the most enlightened energy utility to welcome the introduction of competition in the form of distributed solar energy systems, whether thermal or photovoltaic, in areas where they already have an established distribution network.

Compared with conventional electricity generation, the photovoltaic systems are more suitable for on-site applications, thereby avoiding transmission losses and in some cases offering the possibility of satisfying dc loads with a dc source (eg, radio transmitters or pumps). Widespread diffusion of on-site generators could reverse the trend towards ever-increasing centralised electricity generation. For example, in 1920, 30% of all generating capacity in the USA was on-site: by 1973, the proportion had fallen to 4.2%.

It may also be noted that decentralized systems provide a measure of security to the consumers, since they are less vulnerable to sabotage or direct attack and, of more immediate consequence perhaps, not subject to fuel supply disruptions or labour strikes. There is also the important aspect that where owner and user are the same, on-site generation may encourage deeper respect for energy conservation generally.

Interface between electricity utility companies and photovoltaic systems is crucial. The cost of the photovoltaic system can be substantially reduced if storage requirements are eliminated or made much smaller by connecting the system to the grid. When solar generated electricity is insufficient to meet demand, power can be obtained from the grid and when excess solar generated electricity is available it can be fed into the grid. Utilities may thereby save fuel and transmission costs and may be able to avoid or postpone the installation of additional capacity to meet peak loads, particularly if the peak demand matches the time of peak solar energy availability.

Implementation

The opportunities for the electricity utilities to become major owners and operators of distributed photovoltaic systems need careful consideration. Maintaining the integrity of the grid system with distributed photovoltaic (and possibly also wind) generators will require the introduction of control standards and equipment based on actual experience. Accordingly, it is highly desirable for utilities to implement their own pilot and demonstration programmes to gain experience in this important area.

One attractive solution, technically at least, would be for the utilities to own distributed photovoltaic systems by making roof or land use agreements with their customers. The utility would install, own, operate and maintain the system which would supply electricity to the building or feedback into the grid. Alternatively, the utility could act as the financing authority and provide technical advice on the installation and operation of a system that would be leased to the user.

The question of utility finance is important. If photovoltaic and other renewable energy generators provide an economic alternative to the construction of new conventional generating plant, then it is logical that the utility should finance their construction. The important parameter is marginal cost, since if the decision is left to the consumer, he will base his calculations on the price of electricity he has to pay to the utility. This price is an average for all generating plant, from the oldest and least efficient station to the latest and most efficient. The utility on the other hand must base its calculations for new plant on current construction costs and present and future fuel prices. These calculations are likely to produce higher energy costs, despite improved efficiencies, because of inflation. Distributed photovoltaic systems would on this basis become economic sooner for the utility than they would for the consumer, neglecting any differences in interest rates and tax credits.

The role of the electricity utilities must change if they are to respond to the need to encourage the use of distributed renewable energy generators. They could pay for energy audits, conservation measures, and installation of solar or wind generators, recovering the costs in the rates charged for electricity. In this way, homeowners and industries will be able to avoid the often prohibitively high initial cost and there will be strong incentives for them to participate, to the benefit of all consumers and the nation as a whole.

Buy-back rates

Reference was made in Chapter 3 to the matter of the buy-back rate (ie, the rate paid by a utility for surplus electricity fed into the grid from a photovoltaic generator). One of the most significant features of the Public Utility Regulatory Policies Act (PURPA) of 1978 in the USA is that it directs states to adopt more conservative rate systems that reflect the true cost of the electricity generated and remove unintended or, intended discrimination against on-site generating facilities. PURPA requires utilities to permit small power producers and co-generators to connect to the grid system and feed electricity into it. The utility can be obliged to sell power to and buy power from the operators of private generators at prices that are fair and reasonable.

The price paid by the utility for excess power fed into the grid, the buy-back rate, should not exceed the incremental cost to the utility of alternative electrical energy and should reflect the true 'avoided cost' of providing the equivalent amount of electricity by other means. Consideration must be given to time-of-day, seasonal and interruptible rates, all of which can affect solar energy systems. Buy-back rates and other aspects of the legislation are still subject to debate in the USA but there is no doubt that consideration of the principles will affect the rate of implementation of decentralized photovoltaic and other types of generator. Some of the electricity utilities in Europe already have a similar policy (eg, the CEGB in UK) but it is important for the matter to be given careful attention and clear guidelines established in all European countries, to provide a basis for private investment decisions regarding renewable energy systems such as solar and wind generators.

6.4 Finance and taxes

Having referred to the possibility of utility-provided finance for photovoltaic systems, it is necessary to consider other sources of finance and the influence of taxation rules.

The market for new solar technology is characterized by barriers of high costs and technical uncertainties. Both cost reductions and experience can be gained by encouraging the diffusion of solar systems into the private market and economic incentives can be introduced by governments to assist this development.

In the USA, the Federal Government allows a 40% tax credit to homeowners and 15% tax credit to businesses on the amount spent (up to the first $10000) on solar thermal, photovoltaic, wind or geothermal systems. Other federal solar incentive programmes include increased ceilings on insured mortgages to include solar systems and loan guarantees for solar purchasers. The DOE Office of Appropriate Technology offers grants up to $50000 for the design, development and building of small-scale energy systems for local energy needs. In addition, the National Energy Conservation Policy Act (NECPA) of 1976 allows public utilities to make loans to customers for solar equipment installations.

In addition to the federal incentives, practically every state in the USA has now enacted legislation of some form to encourage the use of solar energy and to implement conservation measures in homes and businesses. Incentives include reduction of excise taxes, property taxes and state income taxes, as well as grants and loan programmes, development of solar-orientated building and zoning codes and information services. In most states, federal and state tax credits can be combined, thereby giving up to 75% credit on the total system cost.

To illustrate the effect of financial incentives and tax credits on a typical photovoltaic system, the effect on life cycle cost from selected examples of financial incentives has been calculated (78). For the base case cost, the following parameters were used:

System life	20 years
Installed cost of system	$10000
Inflation rate	8%
Interest rate	10%
Discount rate	10%
Operation and maintenance expense (percentage of installed cost)	1.5%
Insurance	0.3%
Property tax rate	2%
Sales tax	3%
Personal income tax bracket	25%
Downpayment	10%
Benefit assumed, year after purchase	1 year

This gives a base case Life Cycle Cost of $13758. The life cycle cost variations arising from various examples of financial incentives are shown in Table 6.1. The most significant reductions are produced by the property tax, income tax and substantial interest rate subsidies. For example, if the system could be exempted from property taxes and if finance at 5% could be made available, the Present Value could be reduced to $9425, some 30% less than the base case.

Case Description	Life Cycle Cost, $	Change in Life Cycle Cost, $
Base Case (see text for parameters)	13 758	
Exemption from Sales Tax		
Sales Tax Rate 3%	13 458	300
Sales Tax Rate 5%		500
Interest Subsidy		
Interest Rate $9\frac{1}{2}$%	13 539	219
Interest Rate 5%	11 729	2 029
Interest Free 0%	10 125	3 633
Reduction in Downpayment		
5% Downpayment	13 672	86
No Downpayment	13 587	171
Exclusion from Property Tax Base		
25% Federal Income Tax Bracket	11 454	2 304
Credit on Federal Income Tax		
40% Credit	10 122	3 636
55% Credit	8 758	5 000

Table 6.1 Effect of financial incentives on life-cycle cost
of residential PV systems

Certain tax incentives and direct subsidies are already available in some
European countries to encourage homeowners and businesses to introduce
energy conservation measures and renewable energy systems such as solar
water heaters. Such provisions will certainly be needed to encourage the
introduction of photovoltaic systems, by offsetting the higher interest
payments and higher property taxes that would otherwise fall on the
purchaser. Lending banks, loan institutions, building societies and
insurance companies will also need to be reassured that photovoltaic
systems constitute a reliable and sound investment risk.

6.5 Environmental implications of photovoltaics

Introduction

In comparison with conventional electricity generation methods,
photovoltaics are generally considered to be benign since they are silent
in operation, have no moving parts and produce no toxic fumes. To assess
fully the environmental impact of this energy source, it is necessary
however to consider the entire process required to obtain electricity from
photovoltaics, including mining and refining of raw materials, manufacture
and installation of all components and final disposal at the end of the

useful life of the system.

In some countries, it is necessary to make an environmental impact assessment before proceeding with any major development. Such assessments will almost certainly be required for any large photovoltaic array and thus it is useful to review here the main areas of concern. These are health and safety, nuisance to the general public at large, ecological impact and land usage. Emphasis will be placed on the silicon cell technology since this has reached the level of greatest development and is already being implemented in many places for a wide range of applications.

Health and safety

The health and safety implications of photovoltaics have to be considered for three stages: manufacture, operation and final disposal. For the manufacturing stage, established health and safety regulations from industries such as plastics, electronics, glass and chemicals are directly applicable to all operations needed to make photovoltaic systems. No unique hazards are presented and the operations involved present less risk to life, and in the long term to the environment, than say coal-mining or handling nuclear fuels. Two types of risk are involved here: the normal, reasonably well-defined hazards of the manufacturing and construction process for which established historical data are available; and the less-tangible risk to the environment and to public health associated with say the progressive increase in carbon dioxide content in the atmosphere or the disposal of nuclear waste. Photovoltaic systems involve risks of the first type but none of the second.

Current manufacturing techniques for silicon cells involve the mining of sand or quarzite and its reduction by coke in an electric arc finance to produce metallurgical grade silicon of 98 to 99% purity. This process releases carbon monoxide and particulate silica into the atmosphere. Provided stack heights are appropriate, ground level concentrations can be kept well within safe limits. (Silica is non-toxic but silica dust can produce scarring of the lung tissue causing silicosis and also can build up in the kidneys and other sensitive organs.)

The next stage is to refine further the silicon to produce very pure silicon. In the Siemens process, metallurgical grade silicon is ground and reacted in a fluidized bed reactor with hydrochloric acid. The crude product is distilled to generate pure trichlorosilane ($SiHCl_3$) which is then reacted with hydrogen to deposit pure silicon. Various salts of metal impurities including boron and aluminium chlorides must be disposed of carefully to avoid contamination of ground water, surface water or soil. The pure silicon is then processed into mono- or poly-crystalline ingots or sheets. The environmental problems of silicon production are summarised in Table 6.2.

New processes are being developed for refining silicon to obtain solar-grade material and when details are available these must be analysed for potential health and safety implications.

Large amounts of electrical energy are needed to produce silicon cells. The hazards associated with conventional power generation techniques are well-known and do not need elaboration here.

Process	Primary Hazards	Effects
Mining	Silicon dust inhalation	Silicosis
Electric arc furnace (MG silicon production)	Submicron size silica and dust emission	Silicosis
Siemens process	Hydrochloric acid and silanes to atmosphere, metal chlorides, slag	Poisoning
Crystallization	No significant hazards	-

Table 6.2 Environmental considerations of silicon production

Two other solar cell technologies need to be considered, both of which involving considerably greater toxic risks than silicon. The first is cadmium sulphide/cuprous sulphide solar cells, which offer a promising potential for low-cost photovoltaics. Cadmium and cadmium salts are soluble, although cadmium sulphide is the least soluble form, and therefore care must be taken to avoid any pollution of waterways and the soil. After cadmium sulphide cells have been encapsulated, they constitute a lower environmental exposure risk than from other common substances such as paints and plastics. The manufacturing plant will however require very strict emission controls and monitoring facilities. In addition, fires involving cadmium sulphide arrays are of some concern, since toxic gases would be released, although the problem is no greater than is commonly encountered with burning paintwork.

Gallium arsenide is the other type of cell that may possibly find use, especially in concentrator photovoltaic systems. Gallium and arsenic present similar manufacturing hazards to cadmium. Furthermore, concentrator solar cells are subject to loss-of-coolant accidents with consequent risk of melting and release of toxic gases.

Disposal of photovoltaic arrays would present much the same problems as waste disposal from the manufacturing process. Silicon cells, if not salvaged for reprocessing, may be ground and disposed of in ordinary landfill sites, since silicon is non-toxic and environmentally stable. Special arrangements similar to those employed for other toxic substances will be needed for the disposal of cadmium sulphide or gallium arsenide solar cell arrays.

The operation of photovoltaic arrays should present no particular hazards other than the normal hazards associated with electrical equipment. It is worth stressing that dc circuits provide greater electric shock hazards than ac circuits of similar voltage. A dc voltage of 80V is normally considered a safe upper limit, and it must be remembered that photovoltaic arrays cannot be switched off – the open circuit voltage may be dangerous even in conditions of low irradiance. Short-circuit faults in dc systems also produce more persistant arcs and thus constitute a more serious risk than ac systems of comparable voltage.

Nuisance to the public

Large photovoltaic arrays will constitute a major feature of the landscape, but with careful siting and shielding by hedgerows or fences, the visual impact can be reduced. It will not be impossible to avoid reflections and visual distraction entirely, but these will be no worse than sunlight reflection on expanses of water. Reflector concentrators involve eye hazard areas and care is needed when servicing such systems in sunlight.

Apart from these mainly aesthetic considerations, photovoltaic systems are unlikely to cause any other nuisance. Unless very high voltages are involved, there is little chance of electromagnetic interference with radio, television or radar, although some care will be needed to avoid undesirable reflections. The systems will essentially be noiseless, although some wind noise may be experienced depending on the structural form. Transformers, if incorporated in the system, may give rise to some low frequency hum which can be distracting, unless controlled by suitable design.

Ecological impact

The construction of large photovoltaic arrays on the ground will result in some impact on plant and animal life in the immediate vicinity. No serious disruption will be caused and, if necessary, the design could be such that cows or sheep can continue to occupy the same field. Some market gardeners may find the shading caused by the array a positive benefit and use the area beneath the array to raise high-value crops.

A few other points may be mentioned, although none present any particular problem or hazard. Rainwater run-off from array surfaces has to be controlled to avoid erosion of soil. The arrays also need to be cleaned from time to time and if detergents are used care is needed to ensure the safe disposal of the waste water. Concentrator photovoltaic systems often involve coolant circulation which may be water with anti-fouling and anti-freeze additives or may be a special chemical. In either case, care will need to be taken to ensure safe disposal of coolant in the event of a leak in the system.

Land usage

Land use for photovoltaic systems is dependent on the type of installation. Arrays on buildings involve no extra land requirements. Larger scale systems will however occupy considerable land areas, which in many parts of Europe may present problems since such spare land as there is usually needs to be preserved as a public amenity.

It is clear however that there are areas of derelict land or land of very

low value in many countries which could be employed profitably for
photovoltaic installations. The total land used for a photovoltaic system
may not be much different from conventional energy systems such as nuclear
or coal if the land needed for mining, waste disposal and
transmission/distribution is also included. It has been estimated that a
coal fired power station occupies about $3.8km^2$ /GW and that over its
lifetime it may require fuel derived from strip mining 1300 to 5000ha.
Taking into account transmission and distribution requirements, total land
use may be up to $50km^2$ /GW. Whilst partial restoration is possible, acid
drainage and other problems may continue from the mining and waste tips.
By comparison, a 200MWp photovoltaic plant would require about $3.4km^2$,
equivalent to $16km^2$ /GWp. The peak power of a photovoltaic plant however
needs to be at least five times the power rating of a coal plant for
similar annual output, and thus the land requirements are broadly
comparable.

CHAPTER SEVEN - CONCLUSIONS AND RECOMMENDATIONS

7.1 Photovoltaics: the technology and the prospects

Solar cells and modules

Mono-crystalline silicon solar cells, developed during the 1960's for space
applications, now constitute a proven technology for converting solar
energy directly into electricity at an efficiency of up to 15%. Modern
encapsulation methods should ensure an operating lifetime of at least 20
years, with degradation in performance likely to be no more than 5% over
that period.

Improvements continue to be made in silicon cell technology to achieve
higher efficiency with less waste during production. Poly-crystalline
silicon cell techniques have also advanced from the development stage to
commercial production, with solar cells of 10% efficiency being produced
using cast ingot methods by Wacker-Chemitronic in FR Germany and Semix in
the USA. The search for acceptable methods of producing cheaper
solar-grade silicon to replace the expensive electronic-grade silicon is
now beginning to show results, with at least two factories dedicated to the
production of solar-grade silicon under construction.

Developments of continuous silicon ribbon techniques are close to
commercialization. Mobil Tyco and Westinghouse in the USA have each
started pilot production of their ribbon processes. The CGE group in
France is also developing a silicon promising ribbon process.

In the thin film solar cell area, steady progress continues to be made in
both cadmium sulphide and amorphous silicon technology. Two groups, Nukem
in FR Germany and Photon Power in the USA, are constructing pilot plants to
produce cadmium sulphide cells. SES, also in the USA, has been making
cadmium sulphide cells for some time, but there have been technical
problems and this venture has not been a commercial success. Amorphous
silicon technology is being developed by several groups in Europe, Japan
and the USA. Small amorphous silicon cells are used extensively by
Japanese electronics manufacturers to power watches and calculators. Sanyo
has announced that it is investing $50m with the objective of producing
large area amorphous silicon cells costing $2 to 3/Wp within 2 years. In
the USA substantial additional funding has recently been announced for the
amorphous silicon process being developed by Energy Conversion Devices
(ECD) and commercialization of their technology is said to be imminent. In
Europe, amorphous silicon development is still some way from being ready
for commercial production.

Although a number of concentrator photovoltaic are being developed, the
long-term prospects for this technology, in competition with flat plate
arrays, are uncertain. Apart from certain applications where the
concentrator's higher efficiency and potential for providing thermal as
well as electrical energy can be exploited to the full, the simplicity and
potentially lower cost of flat plate systems render concentrators a less
attractive option for major investment.

Photovoltaic systems

Small stand-alone photovoltaic systems are already commercially viable for applications such as cathodic protection, vaccine refrigerators, telecommunications, lighting units, battery charging and water pumping in remote areas. A wide range of other applications, particularly systems to replace small diesel generators up to about 30kW output, are technically developed and are close to being economically viable, particularly for locations remote from the existing electricity grid.

Larger systems, including systems to operate in conjunction with the electricity grid are still being developed, to find technically appropriate and reliable equipment that will condition the dc power output from the photovoltaic array and convert it to ac power at the required voltage and frequency. Such systems have good prospects for use in industrialized countries and in the developing world. The technology for complete systems is being developed through the design and construction of many pilot plants and demonstration projects in Europe, the USA and elsewhere.

One of the keys to improved cost-effectiveness is to make the maximum use of the energy available from the array. The theoretical and practical problems are now better understood, and components and subsystems are now under development specifically for use as part of photovoltaic systems.

Costs of photovoltaic modules and the associated costs of complete systems have been steadily falling as production volume increases and new techniques are introduced. Modules that in 1976 cost about ECU 30/Wp for large volume orders now cost less than ECU 7/Wp (1980 ECU). Complete systems today cost between ECU 15 to 35/Wp, depending on size and application. It is confidently expected that module prices will continue to fall, significant reductions being likely in the next two or three years following the introduction of highly automated large-scale production facilities, coupled with the availability of lower-cost silicon material.

The expectation now is that module prices will be down to about ECU 2.80/Wp by 1986 and fall further, to ECU 2.00/Wp or lower by 1990. Further cost reduction is dependent on advanced techniques and the introduction of thin film solar cells which are still in the development stage; ultimately modules costing only ECU 0.15/Wp may be possible, with system costs in the range ECU 1.5-2.5/Wp.

The electrical output of a photovoltaic system is in general directly proportional to the amount of light energy received by the array. For stand-alone systems energy storage can be included in the system. Other systems may rely upon the electricity grid for back-up where this is available. Larger grid-connected systems or central generating plants need careful consideration of their impact upon the whole grid system. Conventional electricity generators are 'firm', in that they can operate as and when required, with low risk of failure, given the necessary fuel supply, whereas solar generators are 'non-firm', unless they are combined with a substantial amount of energy storage. Solar generators connected to a network do nevertheless offer some capacity credit, in that they make a contribution to the total installed generating capacity required to meet total demand on the network, as well as saving fuel, and in this respect solar generators in general offer higher capacity credit than wind generators. Further capacity credit studies are needed for large grid networks, taking into account the additional benefit arriving from distributed solar generators with associated reduced risk of simultaneous cloud cover affecting output. A key parameter in these studies will be the reliability factor, or the risk accepted by the utilities of being unable to meet peak demand. Consumers in most industrialised countries have become accustomed to a very high level of reliability from grid supplied electricity, but lower reliability levels (and consequently lower energy prices) may be acceptable for some networks.

In addition to capacity credit and fuel saving potential, distributed photovoltaic generators offer utilities other advantages, such as pollution-free operation, low maintenance requirements, less vulnerability to disruptions arising from social or technical causes and reduced transmission and distribution costs. All such factors will need to be considered by utilities in their planning studies.

Markets for photovoltaics

Total world sales of photovoltaics amounted to about 4MWp in 1980 and probably exceeded 6MWp in 1981, with 5MWp in the United States, 0.8MWp in the European Community and the balance in Japan and elsewhere. Production in 1982 is expected to reach 7.5MWp in the United States, exceed 2MWp in the European Community and be about 2MWp in Japan. Until now and for the immediate future, purchases supported by governments for pilot and demonstration projects and as part of overseas aid to developing countries constitute an important part of the market. Commercial sales are however steadily growing in the applications that are economically viable.

If costs of photovoltaic modules and systems continue to fall as anticipated, a large market will develop in Europe – first for small stand-alone systems and later, as the price falls further, for larger stand-alone systems to supply villages, islands and remote technical installations. By 1990, the European market could reach 20MWp per year, assuming systems are available at less than about ECU 5.00/Wp and that the industry is given suitable encouragement and support, as discussed later in this chapter.

With the introduction of low cost silicon cells or thin film solar cells, enabling systems to be marketed at less than about ECU 3.00/Wp, the prospects for residential photovoltaic systems connected to the grid are much enhanced. Given full co-operation by the electricity utilities, rapid penetration into this market can be anticipated in the late 1980's, particularly in the southern European countries which have higher solar insolation levels. Later, when very low cost systems, of the order of ECU 1.5-2.5/Wp become available, the service, commercial, institutional and industrial sectors will open up and become sizeable markets. The total market in Europe could then reach over 1000MWp per year by the year 2000, and continue to grow at over 50% per year. By the year 2025, it is conceivable that in Europe there will be over 200000MWp of photovoltaic systems, most of which will be grid connected, generating about 10% of the total electricity supply in Europe.

If low cost systems do not become available, due to unforeseen technical problems or lack of official support, then grid connected applications will not in general be economic and the European market will then be confined to stand—alone systems for remote locations. Saturation would be reached in the 1990's at about 50MWp per year.

The worldwide market will be considerably greater ·than the European market, presenting European manufacturers with important opportunities for exports and joint ventures. Most developing countries will wish to install their own photovoltaic manufacturing facilities and, to do this, they will look to the USA, Europe and Japan for the necessary technology, finance and experience. Some developing countries, notably India, Brazil and Singapore, are already developing their own photovoltaic capability as far as possible independent of other countries.

Photovoltaic systems for stand—alone devices such as pumps (both for irrigation and water supplies) and for refrigeration plant will be needed in large numbers. Electrification of the many millions of villages at present without power could proceed with photovoltaic systems at a very rapid pace, given the necessary finance from the international aid and banking agencies.

The domestic market in the USA is also seen as being potentially very large, particularly for grid—connected residential systems and larger systems for the service, commercial, institutional and industrial sectors. In time, multi—megawatt photovoltaic systems could be installed by utilities for central generation.

Taking all these aspects into consideration,the total world market for photovoltaics could reach 400MWp per year by 1990 and 10000MWp per year by the year 2000, well on the way to becoming a significant component of the total energy supply. If the lowest cost targets for systems are not achieved, then the market will be lower but still substantial, at an estimated 100MWp per year in 1990 and 200MWp per year by the year 2000. The total annual sales of photovoltaic systems, based on these market projections, would be in the range ECU 600 to 1200 million in 1990 and ECU 700 to 20000 million in the year 2000. These projections of the total volume and the total value of photovoltaic sales are referred to as the 'CEC 1982' projections.

If the higher estimates of sales volume are to be achieved, the photovoltaic industry must expand at a rate of 40 to 50% per annum over the next 20 years. Growth rates considerably higher than this have in fact been achieved by the industry in recent years and some reports have suggested that a 100% per annum growth rate could continue for many years (see for example reference 71). Until recently, production targets for the photovoltaic industry were being freely quoted in the USA as 1000MWp per annum by 1988 rising to over 20000MWp per annum by 2000, but with official policies being revised by the Reagan Administration, few are prepared to be so optimistic. The 'CEC 1982' projections for the total volume and the total value of photovoltaic business are considered however to be realistic, the higher estimates indicating what might be achieved if the industry is officially supported and encouraged and the lower estimates indicating what might be achieved without official support.

Projections of this nature are essentially speculative and are dependent on many assumptions. The differential between fossil fuel price inflation and general inflation rates will be a key factor influencing the time when photovoltaic systems break even with alternative generating systems. In recent years this differential inflation rate has exceeded 10% per annum but the general view today is that over the next 10-20 years the differential will not exceed 5% per annum on average. The projections for photovoltaic sales the later years of this century will need to be periodically updated. As photovoltaic technology develops then so will other technologies. There are also a number of significant non-technical issues to be resolved relating to the introduction of distributed photovoltaic systems, which may delay the diffusion of photovoltaic systems even when they are technically and economically viable. There is bound to be a substantial place for photovoltaics, but the total size of the energy market and the proportion supplied by photovoltaics in the year 2000 cannot be precisely determined.

High growth rates in the energy sector are not unprecedented. For example, natural gas production in the European Community (9 countries) rose at an average rate of 15% per year from 1960 to 1979, reaching a peak rate of 28% per year in the period 1965 to 1974. If small stand-alone photovoltaic systems are viewed as being similar to consumer products, then the rate of growth in the sale of television sets indicates what is possible. In 1946, about 5000 Americans possessed television sets. By 1956, the total was about 42 million and by 1966 the number had reached saturation level at about 65 million, an average growth rate of 150% per year from 1946 to 1956.

Photovoltaic manufacturers

There has been in recent years a major government sponsored research and development programme in the USA to develop low-cost photovoltaic systems. The main emphasis to date has been on photovoltaic cells and modules but increasing attention is now being given to complete systems for a wide range of applications. Over 20 major demonstration projects have been constructed and a corresponding number are being built or are planned. As a result of this broad-based development programme, there are now some 10 to 15 companies in the USA well-established in the manufacture and marketing of photovoltaic components and systems, many having affiliates and licensee companies in Europe and elsewhere. There are also several companies with new photovoltaic products that are close to commercialisation.

In Europe, there has also been a substantial photovoltaic research programme and there are now about 7 or 8 major companies plus a similar number of smaller ones currently engaged in the production and marketing of photovoltaic components and systems. There are several recent entrants whose products have yet to be assessed commercially.

Elsewhere in the world a number of countries are developing their own photovoltaic industry notably Japan, India, China and Brazil. Several companies are active in Japan but they have yet to market their products on a large scale internationally. Photovoltaics form part of a major solar energy research and development programme in Japan and it is expected that in time Japanese-made photovoltaic systems will present a major challenge to American and European manufacturers. Although several companies have developed crystalline silicon photovoltaic systems, the main thrust of Japanese research and development is now directed towards thin film amorphous silicon technology. Already small area amorphous silicon cells are being used to power thousands of Japanese watches and calculators and large area cells suitable for terrestrial applications are expected to be in production within one or two years.

The governments of many countries have recognised the potential of photovoltaic systems as a world energy resource. They are funding not only basic research but also pilot and demonstration projects which accelerate development and also serve to strengthen their domestic industries for entry into the commercial markets as they develop.

In 1981 Europe's share of the world market for photovoltaic production was about 12%. As European industry builds up capacity and experience, it should be able to secure a much larger share of the total world market. The CEC-sponsored photovoltaic pilot plants, which will incorporate over 1MWp of photovoltaics, will help the industry gain experience which will be useful both for securing export orders and as preparation for future photovoltaic markets within Europe. European manufacturers should be encouraged to secure at least 30% of the total world market (ie, double their present share). By the year 2000, the photovoltaic industry in Europe could be worth about ECU 6000 million per annum, with further major expansion anticipated beyond that.

Non-technical implications

Cost considerations apart, the main non-technical obstacles in Europe and elsewhere to the widespread introduction of photovoltaic systems are:

- Reluctance of potential customers to invest in relatively new, un-proven technology;

- Absence of established codes and standards, and doubts regarding warranty terms;

- Planning restrictions on the installation of solar arrays, and doubts regarding long-term solar access;

- Reluctance of electricity utilities to co-operate fully with owners of private generators;

- Lack of suitable financial encouragements such as low-cost finance and tax credits.

Some of these problems will become of decreasing significance as more photovoltaic systems are installed which perform well and are well publicised. Books and articles are already appearing that will help the public understand the technology and appreciate the potential of photovoltaics (see for example references 71, 79 and 80). Public information programmes sponsored by governments and the photovoltaic industry are needed to help potential customers assess for themselves the potential benefits of having a photovoltaic system. Policy guidelines for planning authorities and the electricity utilities regarding solar energy systems, whether photovoltaic or thermal, will reduce areas of potential conflicts of interest and to encourage the introduction of renewable energy systems that are deemed to be in the national interest.

Some countries, notably the USA, have already instituted programmes to provide technical advice and financial aid to individuals or businesses wishing to install solar energy systems. The possibility of the electricity utilities financing the construction of photovoltaic generators that are connected to the grid is attractive and moreover would make it easier for the utility to regulate the terms under which such systems are operated. Whatever the source of finance, it is important from the viewpoint of the overall economy of the country concerned that economic analyses of competing energy technologies are carried out. Costs that reflect as closely as possible the true resource costs should be used and not costs that are weighted one way or the other by discounts or taxes.

7.2 Recommendations for a Photovoltaic Action Plan in the European Community

Introduction

With photovoltaics forecast to be a significant energy resource not only for the developing world but also for Europe and other industrialised countries, it is important that both industry and the general public are properly prepared to take full advantage of the potential offered by the new technology. Governments need to be aware of the prospects so that due allowance is made in their industrial and energy sector strategies. International aid agencies need information on the availability of proven photovoltaic systems that can make significant contributions to raising living standards in developing countries. Electricity utilities should be encouraged to plan now for the day when private individuals and businesses will want to connect photovoltaic generators to the grid. And, most important, the man in the street needs to know about a renewable energy technology that can not only power spectacular one-off enterprises such as a one-man plane that flew across the English Channel in 1981 but will one day be a useful part of every day life, as familiar as his central heating system or washing machine.

There is evidence that a healthy level of interest in photovoltaics is being shown by governments and industries in the European Community. Equally it is clear that other industrialised nations are developing their capabilities and preparing for the day when photovoltaics becomes an important market sector for exports as well as in the home market. If European industry is to compete in these markets, the stimulation of interest and demand must continue until photovoltaics are manufactured and sold commercially in large quantities.

Since 1975, the Commission of the European Communities has been supporting photovoltaic research and development by providing up to 50% financial support for approved projects by commercial companies, research organisations and universitites. The Commission also undertakes direct research and development, mainly in the field of testing and qualification of photovoltaic cells and modules, at its Joint Research Centre at Ispra, Italy. Recently, a major programme involving 15 photovoltaic pilot plants ranging from 30kWp to 300kWp has been initiated by the Commission in the member countries in order to develop the technology for a wide range of applications. There will also be at least three smaller pilot plants of about 2.5kwp each, designed to develop systems suitable for residential applications. All these pilot plants are due to be in operation by the end of 1983.

It is now necessary to consider how best the CEC can continue to support the development of technically sound and low cost photovoltaic systems and what other actions are needed to ensure that best advantage is taken by member countries of the opportunities presented by this new technology.

A scenario for implementation

Photovoltaics is a new technology which shows great promise. Much has already been achieved and it is now clear that there are no major technical obstacles that would prevent photovoltaic systems becoming a viable energy resource that could be implemented on a large scale. Unfortunately, a purely economic approach to diffusion of a new technology reveals several paradoxes that hinder implementation. The problem has been well stated by Katzman as follows (81):

> On the demand side, the timing of initial adoption and widespread diffusion (of a new technology) depend upon such factors as profitability, the size of the initial commitment and the rate of replacement or addition to the capital stock. On the supply side, manufacturers devote their inventive and productive efforts to those applications where they perceive the markets to be profitable. If the perceived market is small and unprofitable, potential producers of innovations will devote few resources to new product development, and a technically feasible innovation may not come into being. Consequently, that small subset of potential innovations that are actually developed, produced and marketed is determined by anticipated market demand.

This is the classic chicken-and-egg situation: new technology such as photovoltaics will not reach large-scale, low cost commercialization unless a market is perceived by the producer, but the market will not exist unless a potentially superior technology is actually cheaper or better than its alternatives. Manufacturers learn to make a cheaper product as output increases, but output cannot increase substantially unless the product is sufficiently attractive in terms of price and quality to the potential customers for them to take the risk of making investments in new technology. The problem is particularly acute for high capital cost/low running cost systems such as photovoltaics.

Another paradox is that, even when economic break-even cost levels have been reached, potential customers may still decide to wait several years in anticipation of further reduction in system costs. The achievement of this further reduction in system costs is however largely dependent on steadily increasing sales today, not potentially larger sales tomorrow.

How can this vicious circle be broken? One solution is for manufacturers to set prices for photovoltaic systems at less than actual costs ('forward pricing'), in anticipation of large sales and associated rapid progress up the learning curve. This clearly involves much risk for the manufacturer and one which few if any are prepared to take today to any significant extent.

The only alternative would appear to be for governments to provide the necessary support and turn what is technically feasible into an established fact. If it is accepted that photovoltaics provide a potentially viable and desirable energy resource, suitable for widespread implementation alongside (or, in places, instead of) existing energy resources, then there would apear to be a good justification for direct action to encourage the development of the industry both on the supply and on the demand sides, for the benefit of all. Energy is central to the economy of every nation and has long been held to be too important to be left to free market forces. Governments have in consequence become deeply involved in the development and regulation of all major energy resources, including the oil, coal, gas and nuclear industries. Has the time now come for governments to intervene in a positive way in the development and regulation of the solar photovoltaic industry? There would appear to be sound reasons for so doing, since photovoltaics can now be considered as a practicable and viable energy resource offering at least three important benefits:

- a strategic benefit: reduced dependence on oil, coal and nuclear power for electricity generation

- an operational benefit: pollution-free equipment giving long life with low maintenance

- an employment benefit: creation of new industries for the manufacture of hardware and for the design, marketing and installation of systems.

In Europe, a coordinated Action Plan for the implementation of photovoltaics as a new energy resource of major significance could take the form of a phased programme as follows:

Phase 1 Commitment Phase (1982-1983)

The European Community makes a commitment to photovoltaics and establishes a programme for implementation.

Phase 2 Development Phase (1983-1988)

Three main activities sponsored and supported by the European Community:

i)   Research and development, directed towards low-cost photovoltaic materials components and systems.

ii)   Industry build-up, through the construction of large integrated plants for the manufacture of materials, components and systems.

iii) Market development, in Europe and for export, through demonstration plants, public education and technical and economic studies for various applications.

Phase 3 Implementation Phase (1988 onwards)

Financial support from a wide range of sources organised on a scale comparable to that provided for other energy technologies to enable photovoltaic systems to be widely installed by utilities, private individuals and the service, commercial, institutional and industrial sectors.

Phase 1 is largely dependent on political decisions, for which those concerned need to be as fully informed as possible. The three main activities included in Phase 2 are discussed in turn below.

Research and development

The Commission of the European Communities is currently contributing approximately ECU 16 million for over 50 photovoltaic research and development projects (Appendix A). The topics covered are:

Silicon cells:	17 projects	
Alternative cells:	9 projects	
System studies:	3 projects	
Concentration:	5 projects	
Pilot plants:	16 large projects plus at least	
	5 small projects	

All of these projects are scheduled to be completed by June 1983, when a new programme is expected to be introduced.

A progress report on each project other than the pilot plants was presented by the organisations concerned at a Contractors Meeting held in Sorrento, Italy, 23-24 March 1981. The results presented were generally encouraging although for some projects it was too soon for any firm impression to be formed. It is clear however that there is still much useful work to be done to refine techniques and processes, develop new materials, components and systems and generally advance the state-of-the-art.

A series of design review meetings for the pilot plant programme have been held in 1981, culminating in a Final Design Review conference 30 November - 2 December 1981 held in Brussels (72). The 16 projects cover a wide range of applications and involve a rich variety of components and control philosophies. Several plants involve features that have not previously been tried elsewhere, which adds to the value of the programme.

In view of the promising prospects for photovoltaics, a strengthened research and development programme would be justified starting in June 1983, with a considerably enhanced budget. For the five year period 1983-1988, a total contribution by the Commission of the European Communities of ECU 50 million is suggested. The programme should remain broadly based, covering a similar range of topics as the current programme, but the emphasis should be changed to reflect the priorities as now perceived in the areas of cell technology, sub-system components and system development. To ensure that the necessary work on the priority topics is carried out, it may be necessary to relax the present rules limiting the maximum contribution by the Commission for approved projects to 50%.

In the search for low-cost photovoltaic cells, five topics stand out as needing co-ordinated and concentrated efforts, namely:

- Solar grade silicon

    Union Carbide, Semix and other organisations in the USA have announced that they have developed new lower-cost processes for the manufacture of solar-grade silicon and are building large plants for its manufacture.

If the European photovoltaic industry is not to be placed at a serious disadvantage, increased efforts are needed to identify which impurities can be tolerated in solar grade silicon and at what concentrations. At the same time the most appropriate production processes have to be identified. There are several groups working to this end in Europe and their efforts should be encouraged and supported.

– Thin film amorphous silicon

The development of amorphous silicon solar cells of high efficiency and durability must constitute one of the best long–term hopes for achieving low cost photovoltaics. Considerably increased support for research and development efforts to this end is needed.

– Thin film cadmium sulphide

Although solar cells based on cadmium sulphide are on the point of being commercially available, continued research and development is needed to improve efficiency and, equally important, extend the useful life of such cells.

– Ribbon silicon techniques

Continued support for the development of ribbon silicon techniques is needed since success in this field will open up the prospects for low–cost crystalline silicon cells, with their known advantages of long–term stability and non-toxicity.

– Silicon wafer cutting

At present mono– and poly crystalline silicon wafers are usually sawn one at a time from the ingots. A renewed effort is needed to develop efficient multi–blade or multi–wire cutting devices to speed up production and reduce wastage.

Turning now to component and sub–system development, three topics should have priority:

– Energy storage systems

The high cost, limited life and relatively poor efficiency of the present generation of batteries, even of those types most suitable for photovoltaic applications. A major research and development effort is needed to find more cost–effective energy storage systems. Studies are needed, followed where appropriate by hardware development, of various systems including new types of battery, pumped storage, compressed air, flywheels and hydrogen combined with fuel cells. Much of this work will also be of benefit to other energy technologies such as wind generators.

– Array structures

At present, array support structures are relatively inexpensive compared with the cost of the photovoltaic modules. As module prices fall it will become increasingly important to develop low–cost yet long–lasting array support structures for ground mounted arrays. It will be equally important to develop appropriate solutions for roof mounted arrays, in particular investigating the practicality of using the array itself as weather protection, thereby saving on overall costs.

- Power conditioning and control systems

The present series of pilot plants will provide much useful experience in the design and operation of power conditioning and control systems, including dc-dc converters and maximum power point trackers, dc-ac inverters, battery charge regulators and micro-processor based control systems. The next step is for the more successful devices to be further improved with a view to high volume production of packaged units.

The third area for research and development is that relating to complete systems. Detailed studies are first necesssary, the objective being to identify designs that are the most cost effective for different applications and site conditions. Later, the more promising proposals should be built and tested as part of a pilot plant programme. The following applications are suggested for design development leading to pilot plant construction:

- residential grid-connected systems in the 3 to 10kWp range

- central utility system of 1MWp

- pumping systems for deep boreholes

- pumped storage systems

- various cooling applications, for ice making, cold stores and air conditioning

- hybrid wind/photovoltaic systems

- systems for specific industrial applications, especially those that can use dc power such as electroplating, electrolysis and radio transmitters.

Studies are also needed to investigate the implications for grid networks of connecting substantial amounts of photovoltaic generators, with particular reference to capacity credit and the optimum proportion of photovoltaic to conventional generating capacity. The control and protection of grid systems incorporating many photovoltaic generators are also matters that need to be studied.

More research should be carried out to identify failure mechanisms, reliability and lifetime of photovoltaic sytems, to establish confidence in the claims of manufacturers that their sytems will last at least 20 years. This aspect is very important for economic analyses of alternative generating systems.

There is one other area that calls for increased research and development effort in Europe, and that is in connection with the production equipment used in photovoltaic plants. At present, most of this equipment, such as in implanters, cutting tools, furnaces, soldering and encapsulation machines, solar simulators and test equipment comes from the USA or from Japan. Bearing in mind the large growth of the photovoltaic industry foreseen in this report, greater attention should be given to developing

European sources for the production equipment.

Development of the photovoltaic industry

It will be necessary to develop policies for providing finance and other incentives to enable the photovoltaic industry build large production facilities for materials, cells, modules and sub-systems. An integrated factory of this nature with a capacity of 100MWp per year may cost of the order ECU 80-150 million. At least five factories of this size would be a reasonable objective for the five year Phase 2 plan, involving a total investment of ECU 400-750 million. The financial support might take the form of low interest industrial development credit, plus tax concessions for the initial years of operation. Factory sites could no doubt be found in existing designated industrial development areas.

Market development

Three principal activities are needed for the development of the market for photovoltaics, namely: demonstrations of proven systems; institutional framework; and finance. Taking these in order, it is clear that many demonstration plants covering a wide range of applications will be needed to help potential purchasers assess the technical and economic implications. No doubt many of the present series of European Community pilot plants will in time become in effect demonstration plants, and it should not be overlooked that there are already four photovoltaic demonstration projects in operation in Europe (see Appendix D). Further demonstration plants will be needed as and when pilot plant experience indicates that the technology can now be considered as being proven.

The institutional framework needs to be established through such matters as development of codes and standards, information and training activities, guidance for planning authorities and setting up procedures for obtaining finance.

The Commission of the European Communities has already made a start in developing standards for photovoltaic performance measurements and module construction. A logical next step would be to produce a design guide or handbook to provide a basis for the design and use of photovoltaic systems, generally along the lines of the document issued in 1981 by the Solar Energy Research Institute in the USA entitled 'Interim Performance Criteria for Photovoltaic Energy Systems', although a more comprehensive work is envisaged. The following subjects would be covered:

- use of solar radiation data

- design of different types of system

- data on photovoltaic components, batteries, inverters and other balance-of-system items

- performance evaluation

- economic analysis methods

- specification of systems and components

- performance guarantee requirements

- materials and workmanship

- warranty provisions.

The design guide could contain detailed information on several of the current series of pilot plants, to be used as reference material.

Wherever possible, standards and codes should be harmonized between Europe and the USA, whilst the technology is still young. This will help to avoid the fragmentation of the world market and facilitate international co-operation.

Regarding existing planning regulations, building codes and electrical construction standards applicable in European countries, special studies are required to identify where these need amendments or additions to cover the special problems and requirements of photovoltaic installations. This activity is of considerable importance if the introduction of photovoltaic systems in Europe is not to be hindered by inappropriate codes and regulations. The question of solar access and associated legal aspects also needs to be clarified with reference to existing laws and recommendations put forward for appropriate changes.

Much can be done to increase public awareness of the potential of photovoltaics. General public awareness will reduce any resistance to the new technology and will encourage more private individuals and industrialists to take an active role. To be prepared for the anticipated future demand, training schemes for professionals and technicians need to be encouraged or sponsored. Awareness of photovoltaic technology should be introduced into school and college curricula as a precursor to the training of the large numbers of skilled labour that will be needed in future decades to work in the industry.

In the short term development of the photovoltaic industry is assisted by the exchanges of knowledge arising from technical journals, specialist conferences and workshops. Sponsorship of these by the CEC should be continued.

The third aspect of market development is finance for prospective purchasers. Financial assistance can take a number of forms. The simplest would be low cost loans and schemes involving tax concessions of one type or another. Capital grants for approved installations is another approach that could be followed. Further studies would be needed to identify the preferred options for individual countries.

The cost of these market development activities may be considered separately. The demonstration plants could be supported in a way similar to the present series of pilot plants, with the European Community providing about one third of the total costs. A budget of some ECU 200 million should on this basis enable over 100 demonstration projects involving a total of about 4MWp of photovoltaics to be installed in the period 1983-1988.

The development of the institutional framework as described above would be relatively inexpensive, needing no more than ECU 5 million at the most.

The provision of financial assistance would require funding of at least five hundred million ECU to have significant impact. It would probably be necessary to establish separate funding agencies in each country, perhaps covering other renewable energy systems as well, such as wind energy converters.

Summary of recommendations

For ease of reference, the various recommendations for Phase 2 of the proposed Photovoltaic Action Plan in the European Community are summarized below:

i)   Research and development

A broad-based programme but with special emphasis on the following priority topics:

Cells:

- Solar grade silicon
- Thin film amorphous silicon cells
- thin film cadmium sulphide cells
- Ribbon silicon techniques
- Silicon wafer cutting

Components and subsystems:

- Energy storage systems
- Array structures (ground and roof mounted)
- Power conditioning equipment

Systems:

- Studies and pilot plants for various applications including residential grid-connected systems, large central utility systems, wind/photovoltaic hybrid systems, etc.

- Studies of the implications of integrating distributed photovoltaic generators into large grid networks.

Production equipment:

- Support for the development of European sources for the main items of production and test equipment needed for an expanding photovoltaic industry.

A total budget for the contribution by the Commission of the European Communities of ECU 50 milllion for the period 1983-1988 is proposed.

ii)   Development of the photovoltaic industry

Encourage the building of at least five integrated production facilities each with output capacity of 100MWp per annum of materials, cells, modules and sub-systems. Finance to be provided to assist in a total investment of ECU 400-750 million.

iii) Market development

Three main activities are proposed:

- About 100 demonstrations of proven technology for various applications, with financial contribution from the Commission of the European Communities amounting to about ECU 20 million.

- Institutional framework to be established through such means as codes and standards, educational programmes and technical journals and conferences. A contribution by the Commission of about ECU 5 million is suggested for these activities.

- Financial assistance to prospective purchasers of photovoltaic systems through low-cost loans, capital grants and tax concessions. A fund of at least ECU 500 million would be needed, distributed among the member countries of the Community.

7.3 Conclusion

This report presents today's view of tomorrow. The one certainty is that amongst the multitude of factors affecting the future of photovoltaics some at least will change. Not only will photovoltaic technology develop, but so will alternative energy systems. It will therefore be necessary to make regular reviews of the technology and prospects for photovoltaics and re-direct activities accordingly. All signs at present are that photovoltaics will in time become a major contributor to the world's energy supplies.

Photovoltaic systems provide a real opportunity for both developing countries and industrialized nations to increase their energy supplies without increasing their dependence on finite reserves of fossil fuel. The present high price of photovoltaic systems renders them economic only for the more remote locations where the cost of alternatives is also high and maintenance presents difficulties. Given apropriate encouragement and support, the photovoltaic industry will be able to introduce improved techniques and higher volume production, which will in turn ensure that the price of photovoltaic systems continues to fall. Progressive reduction in photovoltaic system costs coupled with the anticipated continued real increase in conventional energy prices will result in a steadily increase in the range of economic applications. The market will grow accordingly, to levels comparable with those for existing energy systems.

Photovoltaics provide a safe, non-polluting and renewable source of energy. They operate whenever there is light and could, when available at low enough prices, provide a major energy resource first for the southern European countries and then in time for the central and northern regions. Continued leadership and increased financial support from the Commission of the European Communities for photovoltaics would be fully justified to ensure that Europe reaps the full benefit of this new technology that has so many attractive features. Continued research and development is needed to find the best ways to reduce costs and the industry will need support for the construction of large manufacturing units. Photovoltaic markets need to be developed, by informing potential purchasers of the benefits offered by photovoltaics and through the provision of low-cost finance and tax incentives. The systems that are already developed and proven need wide demonstration both in Europe and overseas. With the assistance of the European Development Fund, photovoltaic systems can be provided to help developing countries raise food production and improve living standards in the rural areas. No other energy technology offers such an attractive prospect for easing the world's energy problems. Photovoltaic power is emerging from the research and development phase and is becoming a real energy option — what is now required is commitment to the principle at the highest level and a programme for its implementation. Photovoltaic power is undoubtedly a viable energy technology for Europe.

REFERENCES

1. E Becquerel : On electric effects under the influence of solar radiation; Compt. Rend. 1839, vol 9.

2. Standard procedures for terrestrial photovoltaic performance measurements - specification No. 101, Commission of the European Communities Joint Research Centre Ispra Italy, 1980 (EUR 6423 EN).

3. F C Treble : Solar cells; IEE Proc Vol. 127, Pt A, No 8, Nov 1980, p.505.

4. G F Fiegl and A C Bonora : Low-cost monocrystalline silicon sheet fabrication for solar cells by advanced ingot technology; Proc. 14th IEEE PV Specialists Conference, San Diego, 1980, p.303.

5. F Schmid, M Basaran and C P Khattak : Directional solidification of M G silicon by heat exchanger method (HEM) for photovoltaic applications; Proc. 3rd EC PV Solar Energy Conference, Cannes, 1980, p.252.

6. J Lindmayer and Z Putney : Semicrystalline material from metallurgical grade silicon; Proc. 15th IEEE PV Specialists Conference, Orlando, 1981, p.572.

7. E Sirtl : The Wacker approach to low-cost silicon material technology; Proc. 3rd EC PV Solar Energy Conference, Cannes, 1980, p.236.

8. J Fally and C Guenel : Study of the elaboration of semicrystalline silicon ingots; Proc. 3rd EC PV Solar Energy Conference, Cannes, 1980, p.598.

9. C S Duncan, R G Seidensticker, J P McHugh, R H Hopkins, M E Skutch, J M Driggers and F E Hill : Development of processes for the production of low-cost silicon dendritic web for solar cells; Proc. 14th IEEE PV Specialists Conference, San Diego, 1980, p.25.

10. J P Kalejs, B M Mackintosh, E M Sachs and F V Wald : Progress in the growth of wide silicon ribbons by the EFG technique at high speed using multiple growth stations; Proc. 14th IEEE PV Specialists Conference, San Diego, 1980, p.13.

11. J D Heaps, S B Schuldt, B L Grung, J D Zook and C D Butter : Continuous coating of Silicon-on-Ceramic; Proc. 14th IEEE PV Specialists Conference, San Diego, 1980, p.39.

12. E Fabre and C Belouet : Ribbons and sheets as an alternative to ingots in solar cell technology: Proc. 3rd EC PV Solar Energy Conference, Cannes, 1980, p.244.

13. Low-cost solar array project documentation prepared by JPL for 15th IEEE PV Specialists Conference, Orlando 1981.

14. D E Carlson : The status of amorphous silicon solar cells; Proc. 3rd

EC PV Solar Energy Conference, Cannes, 1980, p.294.

15. S R Ovshinsky and A Madan : A new amorphous silicon based alloy for electronic application; Nature vol 276, 1978 p.482.

16. Y Hamakawa : Recent progress of the amorphous silicon solar cell technology in Japan; International Journal of Solar Energy, Vol 1, Spring 1982.

17. E D Castel and M J Soubeyrand : Investigation of the reliability of sprayed backwall Cuprous Sulfide/Cadmium Sulfide solar cells for terrestrial applications; Proc. 15th IEEE PV Specialists Conference, Orlando 1981, p.772.

18. T J Cumberbatch, R Hill, J Woods et al : Large area cadmium sulphide thin films produced by electophoretic deposition; Proc. Solar World Forum, Brighton 1981.

19. R B Hall, R W Birkwire, J E Phillips and J D Meakin : 10% conversion efficiency in thin film polycrystalline (Cd Zn)S/Cu$_2$S solar cells; Proc. 15th IEEE PV Specialists Conference, Orlando 1981, p.777.

20. Brochure and technical documentation issued by the Institute of Energy Conversion of the University of Delaware, 1981.

21. J C C Fan, C O Bozler and R W McClelland : Thin-film GaAs solar cells; Proc. 15th IEEE PV Specialists Conference, Orlando, 1981, p.666.

22. J G Posa : TI airs details of low-cost solar collector; Electronics, Nov 1980.

23. P G Borden, P E Gregory, O E Moore, L W James and H Vander Plas : A 10-unit dichroic filter spectral splitter module; Proc. 15th IEEE PV Specialists Conference, Orlando, 1981, p.311

25. J J Loferski : Proc. 12th IEEE PV Specialists Conference, Baton Rouge, 1976, p.957.

25. R M Swanson : Recent developments in thermophotovoltaic conversion; Proc. 3rd EC PV Solar Energy Conference, Cannes, 1980, p.1097.

26. G Cheek and R Mertens : MIS silicon solar cells: potential advantages; Proc. 15th IEEE PV Specialists Conference, Orlando 1981, p. 660.

27. K Krebs and E Gianoli : Photovoltaic conversion of concentrated solar radiation; CEC Joint Research Centre, Ispra, Italy, 1978 (EUR 5723 EN).

28. B D Shafer, C B Stillwell and H Togami : Development and evaluation of a low cost photovoltaic concentrator module; Proc. 15th IEEE PV Specialists Conference, Orlando 1981, p.317.

29. E R Hoover : Comparative analysis of combined flat-plate PVT collectors with separate photovoltaic and thermal collectors;

Proc. 15th IEEE PV Specialists Conference, Orlando 1981, p.732.

30. Varta Unipower Solar-electric systems : Battery Design Guide; Varta Batterie AG, March 1981. See also H Willmes : Battery systems for energy supply in photovoltaic power plants, Proc. EC PV Pilot Plants Final Design Review Meeting, Brussels 30 Nov-2 Dec 1981.

31. P O Jarvinen, B L Brench and N E Rasmussen : Performance characteristics of solar photovoltaic flywheel storage systems; Proc. 15th IEEE PV Specialists Conference, Orlando 1981, p.636.

32. R Dahlberg : The replacement of fossil fuels by hydrogen; paper presented at Photovoltaics Development Conference sponsored by Monegon Ltd, Geneva, 1981.

33. Small-scale solar powered irrigation pumping systems : report on Phase I of UNDP Project GLO/78/004; by Sir William Halcrow and Partners in association with the Intermediate Technology Development Group, for World Bank, 1981.

34. Small-scale solar power irrigation pumping systems : technical and economic review; by Sir William Halcrow and Partners in association with The Intermediate Technology Development Group, for World Bank, 1981.

35. B E Nichols and S J Strong : The Carlisle house: an all-solar electric residence; Proc. 15th IEEE PV Specialists Conference, Orlando 1981, P.1438.

36. SPS Reference System Report, US DOE/NASA, Oct 1978 (DOE/ER 0023).

37. W Palz, A Simon and D Thieriet : Les applications terrestres des générateurs solaires photovoltaiques; Centre National d'Etudes Spatiales - CNES, March 1974 (CNES 109 2952 B).

38. E L Burgess, K L Biringer and D G Schueler : Update of photovoltaics system cost experience for intermediate-sized applications; Proc. 15th IEEE PV Specialists Conference Orlando, 1981,p.1453.

39. R S Gawlik : Displacement of diesel generators in the LDC's - a Monte Carlo approach; Proc. 14th IEEE PV Specialists Conference, San Diego, 1980 p.1054.

40. M Martinez and R Magar : Lifetime and inflation in photovoltaic systems analysis; Proc. 15th IEEE PV Specialists Conference, Orlando 1981, p.273.

41. European Solar Radiation Atlas, Volume I : Global Radiation on Horizontal Surfaces, Commission of the European Communities, W Groesschen-Verlag, 1979.

42. G Jac-Macris : On the distribution of solar energy in Greece; Hypomnemata of the National Observatory of Athens, Series II, No.43, 1976.

43. Measurements for development of solar and aeolic potential of Greece

for energy purposes, Vol II, Public Power Corporation, Athens, Greece, 1981 (in Greek).

44. B Bourges : Les courbes de fréquences cumulées de l'éclairement solaire : analyse statistique et application au calcul des installations solaires; These de Docteur-Ingenieur présentée a L'Université Pierre et Marie Curie, Paris, France, 1980.

45. Unpublished data for Messina, Sicily, kindly supplied by ENEL, 1981.

46. J A Duffie and W A Beckman : Solar Energy Thermal Processes; John Wiley, 1974.

47. C Mustacchi, V Cena and M Rocchi : Stochastic simulation of hourly global radiation sequences; Solar Energy, vol 23, 1979, p.47.

48. Results of tilt angle calculations for Tremiti Islands presented by Adriatica Componenti Eletronici — ACE at Preliminary Design Review of EC photovoltaic pilot plants, Brussels, 1-2 July 1981.

49. Crucial choices for the energy transition : an initial evaluation of some energy R & D strategies for the European Communities; Commission of the European Communities, 1980 (EUR 6610 EN).

50. J Anderer, A McDonald and N Nakicenonvic : Energy in a finite world — paths to a sustainable future; Report by the Energy Systems Program Group of the International Institute for Applied Systems Analysis — IIASA, Ballinger Publishing Co, 1981.

51. Report of Synthesis Group on Energy Technologies included in preparatory material for delegates to United Nations Conference on New and Renewable Sources of Energy (UNERG), Nairobi, 1981.

52. Basic statistics of the Community — Eurostat 1980; Statistical office of the European Communities.

53. Energy statistics yearbook 1980; Statistical Office of the European Communities.

54. G C Manzoni, L Salvaderi and A Taschini : Integration of photovoltaic generation into a large generating system; Proc. 2nd EC PV Solar Energy Conference, Berlin, 1979 p.552.

55. G P Pacati and A Taschini : Utility aspects of photovoltaics in Europe; Proc. 3rd EC PV Solar Energy Conference, Cannes, 1980, p.46.

56. The installation and operational aspects of private generating plant; Engineering Recommendation G26, The Electricity Council, UK, 1975.

57. Notes of guidance for the parallel operation of private generators with Electricity Boards low voltage networks; Engineering Recommendation G47, The Electricity Council, UK, 1981.

58. Y Chevalier, A Haentjens and B Meunier : French public photovoltaic market stand-alone applications; Proc. 15th IEEE PV Specialists Conference, Orlando 1981, p.201.

59. A Gabriel and A W de Ruyter van Steveninck : Photovoltaics in the context of off-grid small power systems; Proc. 3rd EC PV Solar Energy Conference, Cannes, 1980, p.521.

60. G Korn, U Magagnoli, A Taschini and A Vazio : Impiego della conversione fotovoltaica per la fornitura di energia elettrica a nuclei abitati con elevato costo di allacciamento; ENEL Milano, 1981.

61. F Corbellini : Iniziative dell'ENEL per l'utilizzaione dell' energia solare; 3° Mostra Convegno Internazionale, Genova, 1980.

62. F Amman, E Antognazza, L Braicovich and G Panati : Energia Solare : Una Proposta Organica di Politica Industriale nei Comparti Fotovoltaico e Termodinamico; Instituto di Economia delle Fonti di Energia - IEFE, Milan, Italy, May 1980 (draft).

63. SERI photovoltaic venture analysis long-term demand estimation; MIT Energy Laboratory, 1979.

64. Application analysis and photovoltaic system conceptual design for service, commercial, institutional and industrial sectors, Final Report Vols. I and II; prepared for Sandia Laboratories by Research Triangle Institute, Dec 1979 (SAND 79-7020/I and II).

65. V Albergamo and P Bullo : The Delphos project, Proc. 15th IEEE PV Specialists Conference, Orlando, 1981, p.1208.

66. Study on infrastructure considerations for microwave energy ground receiving stations - SPS offshore rectenna siting study in W Europe; prepared for ESTEC by Hydronamik BV, Netherlands, Nov 1980 (ESA CR(P)1411).

67. In-house market studies for small-scale solar powered pumping systems by Sir William Halcrow and Partners in association with The Intermediate Technology Development Group, 1981.

68. In-house market studies for small-scale solar powered refrigerators by Sir William Halcrow and Partners in association with The Intermediate Technology Development Group, 1981.

69. See call for tenders for consultancy services issued by INDOTEC, Dominican Republic, in connection with BID-funded solar-energy development project, August 1981.

70. M T Katzman and R W Matlin : Toward a solar Texas: the promise of photovoltaics; Texas Business Review, vol. 54, No.3, May-June 1980, p.113.

71. P D Maycock and E N Stirewalt : Photovoltaics - Sunlight to electricity in one step; Brick House Publishing Co, Andover MA, USA, 1981.

72. The European Communities Photovoltaic Pilot Plants : Proceedings of Final Design Review Meeting, Brussels, 30 Nov-2 Dec 1981.

73. M Kamimoto and H Hayashi : Sunshine Project solar photovoltaic program and recent activities in Japan; International Journal of Solar Energy, Vol. 1, Spring 1982.

74. K Hay, J D L Harrison, R Hill and T Riaz : A comparison of solar cell production technologies through their economic impact on society; Proc. 15th IEEE PV Specialists Conference, Orlando 1981, p.267.

75. Editorial note on energy payback on page 2 of 'ARCO Solar News,' vol. 1, No. 2, Sept 1981.

76. T L Neff : Social cost factors and the development of photovoltaic energy systems; MIT Energy Laboratory Report No. MIT-EL 79-026, 1979.

77. A S Miller : Legal obstacles to decentralized solar energy technologies; Proc. of Conference on Non-technical Obstacles to the Use of Solar energy, Brussels, 1980, 1980,p.148.

78. D Carmichael, G Noel, J Broehl, J Hagely, M Duchi and H Smail; Institutional issues - major issues and their impact on widespread use of photovoltaic systems; Proc. 15th IEEE PV Specialists Conference, Orlando 1981, p.280.

79. Solar Electricity : making the sun work for you; Monegon Ltd, Gaithersburg, MD, USA, 1981.

80. G C Jain : Development and manufacture of photovoltaic systems in developing countries; National Physical Laboratory, New Delhi, India, March 1981 (for UN University, Tokyo).

81. M T Katzman : Paradoxes in the diffusion of a rapidly advancing technology: the case of solar photovoltaics; Technological Forecasting and Social Change vol. 19, 1981, p.227.

GLOSSARY OF TERMS AND UNITS

This Glossary is in three sections:

I   Photovoltaic terminology
II   General meteorological terminology
III Abbreviations and conversion factors

The definitions are based on the draft recommendations contained in
'Radiation nomenclature – definitions, symbols, units, related quantities'
produced by the CEC Project F ad hoc Working Group, November 1981,
supplemented by the definitions given in 'Standard Procedures for
Terrestrial Photovoltaic Performance Measurements,' published by the CEC
Joint Research Centre, Ispra, 1980.

I   PHOTOVOLTAIC TERMINOLOGY

Solar Cell

The basic photovoltaic device which generates electricity when exposed to
sunlight.

Module

The smallest complete environmentally protected assembly of interconnected
solar cells.

Panel

A group of modules fastened together, pre-assembled and interconnected,
designed to serve as an installable unit in an array.

Array

A mechanically integrated assembly of modules or panels together with
support structure but exclusive of foundation, tracking, thermal control
and other components, as required, to form a dc power producing unit.

Array Sub-Field

A group of solar photovoltaic arrays associated by a distinguishing feature
such as physical arrangement, electrical interconnection or power
conditioning.

Array Field

The aggregate of all solar photovoltaic arrays generating power within a
given system.

Photovoltaic System

An installed aggregate of solar arrays generating power for a given
application. A system may include the following sub-systems:

(a) Photovoltaic array field,

(b) Support foundation,

(c) Power conditioning and control equipment,

(d) Storage,

(e) Active thermal control,

(f) Land, security system and buildings,

(g) Conduit/wiring, and

(h) Instrumentation.

Cell Area

The entire frontal area of the solar cell, including the contact grid.

Module Area

The entire frontal area of the module, including borders, frame and any protruding mounting lugs.

Array Area

The entire frontal area of the array, including intermodule spacing and framework.

Short-Circuit Current (Isc)

The output current of a photovoltaic generator in the short-circuit condition at a particular temperature and irradiance (mA or A).

Open-Circuit Voltage (Voc)

The voltage generated across an unloaded (open) photovoltaic generator at a particular temperature and irradiance, as measured with a voltmeter having an internal resistance of at least $20k\Omega/V$ (mV or V).

Current-Voltage (I-V) Characteristic

The output current of a photovoltaic generator at a particular temperature and irradiance, plotted as a function of output voltage.

Maximum Power

The power at the point on the current-voltage characteristic where the product of current and voltage is a maximum (W).

Conversion Efficiency (η)

The ratio of the maximum power to the product of area and irradiance, expressed as a percentage:

$$\eta = \frac{\text{Maximum Power}}{\text{Area x Irradiance}} \times 100\%$$

Fill Factor (FF)

The ratio of maximum power to the product of open-circuit voltage and short-circuit current:

$$FF = \frac{\text{Maximum Power}}{Voc \times Isc}$$

Spectral Response [$S(\lambda)$]

The short-circuit current density generated by unit irradiance at a particular wavelength (AW^{-1}), plotted as function of wavelength. (The term 'response' is commonly used but strictly correct is 'responsivity'.)

Relative Spectral Response [$S(\lambda)rel$]

The spectral response normalized to unity at wavelength of maximum response.
$S(\lambda)rel = S(\lambda)/S(\lambda)max.$

Time Constant

The time required for a radiometer or photovoltaic generator to attain 63.2% of its steady state value after a step change of irradiance (μS).

Angle of Incidence

The angle between the direct solar beam and the normal to the active surface (degrees).

Reference Solar Cell or Module

A solar cell or module used to measure irradiance and to set simulator irradiance levels in terms of a standard solar spectral energy distribution ('standard sunlight'). It is a cell or module, having essentially the same configuration and relative spectral response as the test cells or modules, which has been calibrated at $25 \pm 2^{\circ}$ C in terms of short-circuit current per unit of 'standard sunlight' irradiance ($AW^{-1}m^{-2}$) by an approved Solar Cell Calibration Agency.

Standard Test Conditions (STC)

Cell junction temperature, as measured at the front contact: $25 \pm 2^{\circ}$ C.

Irradiance, as measured with a reference solar cell: $1000Wm^{-2}$.

Standard solar spectral energy distribution: AM 1.5 direct irradiance normalized to give $1000Wm^{-2}$.

Rated or Peak Power (Wp)

The power output of a photovoltaic generator at a specified operating voltage under Standard Test Conditions (W).

II GENERAL METEOROLOGICAL TERMINOLOGY

Air Mass

The length of path through the Earth's atmosphere traversed by the direct solar beam, expressed as a multiple of the path traversed to a point at sea level with the sun at zenith.

Solar Elevation

The angle between the direct solar beam and the horizontal plane (degrees).

Direct Irradiance (E_{dir})

The radiant power from the sun (and a small area of sky surrounding it, defined by the acceptance angle of the pyrheliometer) incident upon unit surface area (Wm^{-2}).

Diffuse Irradiance (E_{diff})

The radiant power from the sky incident upon unit surface area (Wm^{-2}).

Global Irradiance (E_{glob})

The total solar radiant power incident upon unit area of a horizontal surface (Wm^{-2}).

Global Irradiance = Direct Irradiance (horizontal) + Diffuse Irradiance (horizontal).

Total Irradiance (E_{tot})

The total solar radiant power incident upon unit area of an inclined surface (Wm^{-2}).

Spectral Irradiance (E_λ)

The irradiance (global, direct or diffuse) per unit bandwidth at a particular wavelength ($Wm^{-2} \mu m^{-1}$).

Spectral Photon Irradiance (N_λ)

The photon flux density N_λ at a particular wavelength ($cm^{-2} sec^{-1} \mu m^{-1}$)

N_λ may be calculated from E_λ by

$$N_\lambda = 5.035 \times 10^{14} \lambda \ E_\lambda \quad (\lambda \text{ in } \mu m)$$

Spectral Irradiance Distribution

Spectral irradiance plotted as a function of wavelength.

Direct Insolation

The radiant energy from the sun (and a small area of sky surrounding it, defined by the acceptance angle of the pyrheliometer) incident upon unit surface area during a specified time period (kWh m^{-2} per hour, day, week, month or year, as the case may be).

Diffuse Insolation

The radiant energy from the sky incident upon unit surface area during a specified time period (units as for direct insolation).

Global Insolation

The total solar radiant energy incident upon unit area of a horizontal surface during a specified time period (units as for direct insolation).

Global Insolation = Direct Insolation (horizontal) + Diffuse Insolation (horizontal) for the same time period.

Pyranometer

A radiometer normally used to measure global irradiance (or, with a shade ring or disc, diffuse irradiance) on a horizontal plane. Can also be used at an angle to measure the total irradiance on an inclined plane which in this case includes an element due to radiation reflected from the foreground.

Pyrheliometer (sometimes called a 'Normal Incidence Pyrheliometer' or NIP)

A radiometer, complete with a collimator, used to measure direct irradiance.

Turbidity

The reduced transparency of the atmosphere, caused by absorption and scattering of radiation by solid or liquid particles, other than clouds, held in suspension. As defined by Angstrom, the turbidity of the atmosphere is related to t, the extinction coefficient at a wavelength of 1000nm (normally called the 'turbidity coefficient') and ζ , the wavelength exponent in the expression for the aerosol extinction function:

$$a_{D,\lambda} = t \lambda^{-\zeta}$$

t values less than 0.10 denote a very clear condition, whereas values greater than 0.20 are a distinctly hazy condition. The average value of ζ , which is dependent on the particle size distribution, was assumed by Angstrom to be about 1.3.

Precipitable Water Vapour Content

This quantity is measured by a length in cm which is given by the volume of precipitable water vapour (cm^3) in a vertical column of the atmosphere $1cm^2$ in cross section.

Ozone Content

This quantity is measured by a length in cm which is given by the volume of ozone (cm^3) at STP in a vertical column of the atmosphere $1cm^2$ in cross section.

III ABBREVIATIONS AND CONVERSION FACTORS

CEC   Commission of the European Communities
Cz   Czochralski (silicon crystal growth process)
DOE   US Government Department of Energy
ECU   European Currency Unit
GW   Gigawatt (10^9 Watt)
J   Joule = 1 Watt/sec
kWh   kilowatt hour
MPPT   maximum power point tracker
mtoe   million tonnes oil equivalent (see equivalents below)
PV   photovoltaic
PVT   Photovoltaic/thermal modules or systems
R&D   research and development
SCII   Service, commercial, institutional and
       industrial sectors of the economy
TW   terawatt (10^{12} Watt)
UPS   uninterruptible power supplies
W   watt
Wp   peak Watt

1 mtoe = 11.63 x 10^9kWh = 1.33GWyr = 10^{13}kcal = 4.19 x 10^{16}J

APPENDIX A - CEC PHOTOVOLTAIC RESEARCH PROGRAMME

The total of the CEC contribution (up to 50% of costs) for all approved contracts under Project C, photovoltaic power generation is EUA 3.75 million. All contracts started on 1 July 1980 and most continue until June 1983.

Subject I: Silicon cells

Organisation	Title
Kath. Un. Leuven (B)	Development of new techniques for single crystal silicon solar cell fabrication
RTC La Radiotechnique Complec (F)	Study of mono- or polycrystalline solar cell process, using screen-printing technology.
Laboratoires de Marcoussis (F)	Optimisation of an ion implantation without mass separation - laser annealing technique in order to continuously produce junctions for polycrystalline silicon solar cells.
Lamel (I)	Investigation of potentiality offered by ion implantation as a technique to fabricate high efficiency solar cells.
Technical University of Denmark (DK)	Production of solar cells on the basis of low cost silicon by application of ion implantation, laser annealing, and laser-induced diffusion.
Stichting voor Fundamenteel Onderzoek der Materie (NL)	Optimization of polycrystalline silicon solar cells produced by ion-implantation or deposition and pulsed laser annealing.
Laboratoires de Marcoussis (F)	Design, construction and optimization on the industrial prototype scale of a furnace able to produce polycrystalline silicon ingots as material for solar cells.
Pechiney Ugine Kuhlmann (F)	Fabrication de bandes de silicium en continu pour usage photovoltaique par une nouvelle methode de crystalliation.

Laboratoire d'Electronique et de Physique appliquee	Three-year programme for the study of substrate and growth related problems in continuous polycrystalline silicon layers achieved by the RAD process
France-Photon (SA Moteurs Leroy-Somer) (F)	Implementation of low cost semi-crystalline silicon solar cells and introduction of solar grade poly-silicon
Consortium fur Elektrochem Industrie GmbH (D)	Classification of crystal defects in solar base material with diamond lattice.
RTC La Radiotechnique Compelec (F)	Optimization of processing conditions of solar cells versus physical properties of relatively low cost silicon...
Ansaldo (I)	Introduction of Silso material of Wacker (10 x 10cm) in Ansaldo photovoltaic flat panel production.
Photowatt International (F)	Studies relating to new encapsulation materials
Instituto Guido Donegani (Gruppo Montedison) (I)	Low surface reflecting polymeric materials for photovoltaic encapsulation.
JM Chemie (D)	Encapsulation of photovoltaic solar cell modules.
Resart-Ihm AG (D)	R&D work on the encapsulation of solar cells with improved potting and cover materials.

Subject II: Alternative cells

Organisation	Title
University of Dundee (UK) Max-Planck-Institut (D)	Amorphous silicon photovoltaic junctions produced by gas-phase doping and implantation.
CEA/CENG/LETI (F)	Hydrogenated amorphous silicon photovoltaic generator.
University of Sheffield (UK)	Development of sputtered thin film amorphous silicon solar cells.
Plessey Research Ltd (UK)	Improved amorphous silicon devices.
Universita di Roma (I)	Preparation, study and characteri-

zation of hydrogenated amorphous
silicon for photovoltaic cells.

USTI, Montpellier (F)	Studies to improve the efficiency of
ENSCP, Paris (F)	Cu_2S-CdS spray solar cells.
UHA, Mulhouse (F)	
UAM, Aix-en-Provence (F)	

Ecole Nationale Superieure de	Electrolytical preparation and condi-
Chimie de Paris (F)	tioning of cuprous sulphide.

Thorn-EMI Ltd (UK)	Electrophoreted thin films for low cost solar cells.

Battelle-Institut,	R&D work aimed at the develop-
Frankfurt (D)	ment of a cadmium selenide solar cell for the direct terrestrial trans- formation of solar energy into electrical energy.

Subject III: System studies

Organisation	Title
HOLEC Research (NL)	Optimisation research into a complete photovoltaic generator/consumer appliance system employed for small independent electricity supply sys- tems, deep-water and surface-water pumps and cathodic protection
IDE Industrie Developpe- ment Energie (B)	A power conditioning interface for a photovoltaic mini utility.
Societe Europeenne de Propul- sion (F)	Capteur mixte thermique et photovol- taique a concentration.

Subject IV: Concentration

Organisation	Title
Phoebus (I) CNRS/Pirdes (F)	Test of photovoltaic concentrator Test and demonstration of concen- trating photovoltaic generators Sophocle under mediterranean climatic conditions.
Leonhardt, Addra und Partner Beratende Ingenieure GmbH (D)	Development of concentrator photo- voltaic systems of economic viability using highly concentrating spherical metal membrane glass laminated mirrors for 500W.

Fraunhofer Gesellschaft zur
Forderung der angewandten
Forschung (D)

Solar energy conversion on the basis
of fluorescent planar concentrators.
Set-up of a test collector with
20-30W power output.

Universitat Stuttgart (D)

Holographic thin film system for
multijunction solar cells.

ENEL (I)

High concentration PV 100W module
making use of spectral splitting
and Si-GaAlAs coupled cells.

APPENDIX B - CEC PHOTOVOLTAIC PILOT PLANTS

The total cost of the photovoltaic pilot plants is approximately EUA 30 million of which the CEC contribution is about one third. The contracts started in March 1981 and all plants are scheduled to be completed and in operation by mid 1983.

Project	Peak Power kWp	Site	Application
1	300	Pellworm Island (Germany, FR)	Power supply for a vacation centre, in conjunction with grid
2	100	Kythnos Island (Greece)	Power supply to island network, in conjunction with diesel and wind generators
3	80	Alicudi Island (Italy)	Power supply to island community (120 inhabitants)
4	30	Marchwood near Southampton (United Kingdom)	Power supply to the grid (originally rated at 80kWp)
5	65	Tremiti Islands (Italy)	Water desalination
6	63	Chevetogne (Belgium)	Power for solar heated swimming pool
7	35	Kaw, French Guyana	Power supply to a remote village
8	50	Mont Bouquet (France)	Power supply to TV transmitter of Telediffusion de France
9	50	Nice (France)	Nice airport technical systems
10	50	Fota Island, near Cork (Ireland)	Electricity supply for a dairy farm, in conjunction with grid
11	50	Terschelling Island (Netherlands)	Power supply to a marine training school, in conjunction with grid
12	50	Aghia Roumeli, Crete (Greece)	Electrification of a remote village

13	45	Giglio Island (Italy)	Water disinfection, agricultural coldstore
14	44	Rondulinu Cargese, Corsica (France)	Power supply to dwellings, a dairy and a workshop and for water pumping
15	30	Hoboken (Belgium)	Hydrogen production and water pumping in industry

In addition, there will be at least three projects to be installed at Adrano, Sicily, Italy, alongside the site of the 1MW Eurelios solar thermal generating station project. All systems will be connected to the ENEL grid. The systems at present proposed are as follows:

- 2.5kWp polycrystalline silicon flat plate system from Solaris (Italy)

- 2.5kWp polycrystalline silicon flat plate system from Ansaldo (Italy)

- 2.5kWp tracking flat plate system (Helioman) from MAN (F R Germany)

There are also likely to be at least two smaller systems installed at Adrano in late 1983, one using gallium arsenide cells (ENEL/CISE, Italy) and another using amorphous silicon cells (Siemens, F R Germany).

APPENDIX C - EUROPEAN COMMUNITY OVERSEAS AID PROJECTS INVOLVING PHOTOVOLTAICS

The European Community and its Member States are among the foremost world suppliers of aid to the developing countries in the field of energy co-operation. The means available are as follows:

- Lome Convention I and II in respect of the ACP (Africa, Caribbean, Pacific) countries;

- bilateral agreements with the Mediterranean countries;

- aid to non-associated (Asia, Latin America, non-ACP Africa) developing countries;

- financial support for non-governmental organizations (NGO's).

The projects thus financed involving photovoltaics include the following (as at 15 November 1980):

Subject	Country	Amount allocated (EUA)	Remarks
Lome I			
Electricity supply for a herzian relay	Comoros	200000	in progress
Creation of irrigated perimeters in Logone and Chari provinces (one 5kWp solar pump)	Cameroun	350000	in progress
Mediterranean countries			
Scientific cooperation with the CERS	Syria	763000	in progress
Setting-up of a centre for the application renewable energy sources (EREDO)	Egypt	8000000	under review
Non-associated developing countries			
Solar energy demonstration	Pakistan	1500000	under review

NGO's

Equipping of three boreholes with solar pumps using photovoltaic cells (San district)	Mali	60000	completed
Solar pumps project (1kWp) at Kanel	Senegal	75000	completed
Solar pump project (1.3kWp) at Yangasso	Mali	85000	completed
Solar pump project (1kWp) at Timbuktu	Mali	75000	completed
Solar pump project (9 x 1kWp at Thies	Senegal	250000	completed
Solar pump project (1.3kWp) at Yangasso	Mali	21000	completed
Equipping of a borehole with a solar pump (1.3kWp) at Safolo	Mali	35000	completed
Installation of a solar pump at Gouray	Upper Volta	24000	under review
Production of three solar pumps	Upper Volta	89000	under review
Ten solar pumps using photovoltaic cells	Mali	306000	under review

APPENDIX D - CEC DEMONSTRATION PROJECTS INVOLVING PHOTOVOLTAICS

The following demonstration projects involving photovoltaic systems are supported by the CEC:

Application	Peak Power	Contractor
Electricity for a cold-store for flowers in Calabria, with mirror trough concentrator photovoltaic array	30kWp	Officine Galileo, Italy
Simulation of a photovoltaic grid-connected house	5kWp	University of Gent, Belgium
Electric car project. The batteries are charged in a storage room equipped with photovoltaic panels	2kWp	University of Leuven, Belgium
Microirrigation with a Pompes Guinard water pump near Bordeaux	924Wp	Elf Aquitaine, France